Fundamentals

of Living and Nonliving Universes from Black Hole to Cancer

KAMBIZ AFRASIABI M.D.

PAGE PUBLISHING, INC.
Conneaut Lake, PA

First originally published by Page Publishing 2018

ISBN 978-1-64214-737-7 (hc)
ISBN 978-1-64214-736-0 (digital)

Printed in the United States of America

DEDICATION

"This book is dedicated to my beloved mom, from whom I learned unconditional love."

IN MEMORY OF MARYAM MIRZAKHANI

The late Maryam Mirzakhani was an Iranian mathematician, professor at Stanford University, and the first female to be awarded Fields Medal in mathematics. She was light in the darkness and a role model for generations to come.

ABOUT THE COVER

This is a painting by Master Davood Roostaei, representing the author's general design and view. The author believes that unlocking the invisible locks placed on minds would lead to enlightenment, major discoveries, and exceptional milestones in art, music, and science. For that reason, the author coined the name *Unlocked Mind* for this masterpiece.

PREFACE

From the first record of cancer manifesting as a breast lump in ancient Egypt to today's use of targeted therapy, much has been studied in terms of cancer etiology, diagnosis, and resistance to treatment. In this new era of rapid advancement in technology, cancer research has veered toward gene therapy and immunotherapy, as well as innovative methods of cancer screening. However, our knowledge has continued to remain limited in the grand scheme of the living and nonliving universes. In order to pass through these barriers, we must approach the old problem with a different perspective.

This book is intended for scientists and anyone curious and interested in looking beyond our three-dimensional view of the universe. The goal is to shed light on the inner workings of living and nonliving universes and generate a novel perspective on cancer etiology and treatment.

"Afrasiabi law of spontaneity in living universe, Afrasiabi infinity wall of nonliving universe, and super quantum vector and its applications in future cancer treatment protocols are among the new concepts presented for the first time by the author in this book. In addition, the author also proposes the urgent need to generate a new field of science, which he has coined "Quantum Biology". Quantum biologists are expected to master the laws that govern the

homeostasis of the known non-living universe

(the ocean), as well as the living universe (the

fish). This would enable them to look at the

old problem from a new point of regard."

CHAPTER 1

Introduction and Basic Concepts

True understanding of cancer cell necessitates deep understanding of normal cell. This in turn demands underpinning the birth and evolution of life.

As life itself came into existence in the vast ocean of known universe, this makes understanding the basic principles that govern the birth and homeostasis of universe a prerequisite to all.

The known universe is said to have originated in big bang. Throughout that process, massive amount of energy has come into existence which has also converted to matter.

We are born out of 4 or so percent of the whole matter in the known universe. Ninety-six or so percent of the matter spread in the known universe is dark matter.

The most fundamental law governing the known universe at all levels is the second law of thermodynamics.

By the virtue of this law, the index of disorderliness which correlates with entropy incessantly increases following the birth of the known universe.

The total amount of regular energy in the known universe is said to be zero, which is reasoned by positive energy (force of expansion) counteracting negative energy (force of attraction).

These two exactly cancel each other out. So is the case with the total regular mass in the known universe, simply because $E = mc^2$.

Life is the only machinery on the face of the known universe in which the speed of rise in entropy is the lowest as set by the limits of the second law.

The barrier generated by the cell membrane in the primordial ocean on earth to create the first unit of life some thirty-eight hundred million years ago demanded an energy source to maintain its integrity as well as that of its constituents.

The birth of this unit could be considered the first move against the fast pace of increase in entropy in the surrounding environment.

Glycolysis in the anaerobic prokaryotes and oxidative metabolism of nonoxidative phosphorylation subtype in the aerobic prokaryotes seem to have served that purpose.

With the birth of eukaryotes, oxidative phosphorylation became the dominant source of energy supply.

Eukaryotes not only existed as unicellular organisms but also had the bioenergetics sophistication and advantage to evolve into multicellular organisms.

The simultaneous birth of multicellular era and transmembrane proteins such as G-protein coupled receptors as well as the dominance of oxidative phosphorylation seem to be more than a mere coincidence.

With the convergence of this triad, a more efficient energetics machinery started to handle the energy demands of a much more organized biosystem.

Transmembrane proteins connected extracellular and intracellular compartments. This connection offered the eukaryotes the opportunity of constant surveillance of environmental cues.

GPCRs took over the task of fine and efficient distribution of available energy through their downstream pathways, namely C-AMP and PI3 Kinase.

This could secure transition into a more complex biosystem with significantly less entropy as compared with the prokaryote and unicellular eukaryote era.

In unicellular and multicellular eukaryotes, oxidative phosphorylation of one molecule of glucose could generate thirty-six molecules of ATP. This is an eighteenfold increase in bioenergetics efficiency.

Transformation of a normal cell into a cancer cell is associated with a metabolic shift from oxidative phosphorylation to aerobic glycolysis, which generates four molecules of ATP from one molecule of glucose. This has been well described as Warburg's effect.

Currently, cancer is defined as uncontrolled proliferation of cells associated with disorderly maturation. This is more of a description than a definition.

The new definition of cancer should reemphasize the metabolic derangement of cancer cell as the central event.

One of the hallmarks of cancer is a shift back in time as far as dominance of bioenergetics machinery is concerned.

Cancer cell relies on glycolysis, which is the hallmark of prokaryotes as the main source of energy.

As mentioned earlier, this is also called Warburg effect. In 1956, Warburg discovered that cancer cell relies on aerobic glycolysis as the main means of metabolism.

Consequently, cancer could also be defined as regression in evolution of normal state and its energetics machinery.

With that regression, comes a significant increase in entropy and disrespect to barriers built into the multicellular system

Migration of cancer cell to unchartered territories of other organs could be best defined as an attempt at spreading this regressive energetics move and shift back in time.

Cancer cell becomes blind to its position in its native organ and metastasizes to foreign organs. This is a complex biological process.

It demands sophisticated genetic machinery to penetrate established barriers and the ability to survive in foreign organs and usurping their resources.

In evolutionary biology language, cancer cell obeys the "grow or go" principle. This principle serves the immediate purpose of survival of cancer cell but culminates in demise of the host.

DEFINITION OF ENERGY

Deep understanding of energy and its dynamics in the biouniverse is a prerequisite to the understanding of normal cell homeostasis.

Normal cell depends on fine and well-balanced energy distribution to its critical compartments and maintenance of the highest possible free energy.

This understanding is much needed for underpinning of pathological conditions such as cancer.

Energy is defined differently in different systems. In Albert Einstein's famous equation, $E = mc^2$.

Thus, energy and mass are interconvertible and that clearly necessitates deeper understanding of the nature of matter as well.

In 1873, the American engineer Gibbs defined energy in closed thermodynamic systems with constant temperature and pressure as the force that can do effective work in a nonvolume-enhancing fashion.

However, the truth about the universal nature of energy could be totally different and much more puzzling and elegant.

Grasping a better understanding of the fabric of the known universe might offer us the unique opportunity to come up with such essential understanding.

In this sense, the following example might help us grasp a clearer picture of the nature of matter and energy.

It is said that if we increase the size of a proton to the size of the known universe, the size of the superstrings comprising that proton would be as tall as a tree or around ten feet.

Superstrings are the smallest representation of matter, manifesting themselves as vibrating strings.

In other words, the number of seconds in 13.7 billion years (age of known universe) × 300,000 kilometers per second (speed of light) converted into feet = A, divided by 10 feet is roughly the condensation ratio of energy into matter in the known universe.

This condensation force/CF could be deducted by applying a constant such as K (universal condensation constant) to the condensation ratio. As such, KA/10=CF.

In other words, energy as defined by Albert Einstein has become condensed in a mind-boggling way to convert into an entity that we perceive as matter.

What we have learned as big bang, which presumably has created the known universe, is perhaps the only true universal, however, puzzling force.

The nature of this force, if indeed a force and not progenitor of the force, is not known to scientific community.

This presumable force has generated space-time and all its constituents. It has been said that this force might have represented itself as or translated into Hawking's radiation or energy density of the known universe.

Mathematical calculations have suggested that this energy density was around one with thirty zeros degrees centigrade hot and one with ninety-two zeros grams per cubic centimeter dense.

Hawking's radiation has thus been suggested to have given birth to the known universe.

This puzzling force could have created the whole landscape of the known universe in no time.

Lack of deep understanding of the nature of this puzzling force and the mechanisms that have gone into the creation of the known universe have long plagued the field of physics.

This lack of understanding has ill ramifications that extend into all fields of science.

Because of this puzzling force, which could also be called the true force, the end of known universe was created simultaneously with its birth.

The outer most border of universe could be envisioned as an impermeable wall. I have come to call this wall the infinity wall or Afrasiabi wall.

Beyond this impermeable wall, there is nothing such as space-time that is definable or comprehendible to us.

The reflection of the residual force of big bang off the infinity wall has probably given birth to regular energy, which relates itself to regular matter by Einstein's equation.

This reflection has probably happened because of lack of further capability of the weakened force of big bang to generate more dark matter and dark energy and thus more space-time.

This reflection probably has happened on a platform of dark matter, which is probably the most immediate and fundamental derivative of the force of big bang.

Massive and mind-boggling condensation of the reflected and weakened force of big bang, which could by now be called regular energy, down the dark path of dark matter could have generated clumps of regular matter.

Dark matter constitutes the fabric of the known universe. This fabric is held stable by probable dark energetic vibrations or dark energy.

Like dark matter, dark energy could be another immediate derivative of the force of big bang.

The quantum nature of the known universe also implies that dark energy and dark matter could be the two sides of the same coin.

The process of generation of space-time is pretty much like the way a balloon is inflated and kept stable.

The blowing force could be considered the force of big bang or a mysterious progenitor of the force of big bang.

The molecules of air filling the created balloon or known universe would be the dark matter.

The force keeping the molecules of air or dark matter stable would be the dark energy. The outer-most border of the balloon would be the infinity or Afrasiabi wall.

Condensation of regular energy into super-strings could have led to formation of matter as we know it, in post big bang phase and following reflection off the infinity wall. One could arbitrarily call this phase the big condensation phase.

Thus, big bang itself might not have generated regular matter as is currently conceived and taught at academic centers.

Big bang as we know it is an expansive force or the byproduct of an enigmatic expansive force while generation of regular matter demands condensation force.

However, big bang, through generation of space-time and its skeleton/dark matter, could clearly have generated the platform for creation of regular matter in post big bang phase or big condensation phase.

Consequently, it seems quite likely that big bang could have created space-time and dark matter as the fabric of space-time simultaneously.

In addition, it could have also directly, and in part, converted into dark energy, which could be considered the stabilizing force of dark matter as well.

Alternatively, dark energy could indeed be nothing but quantum fluctuations of dark matter in the quantum universe.

In other words, dark matter and dark energy could be the immediate quantum derivatives of big bang, both born simultaneously with the birth of space-time out of near zero geometry and near zero space-time.

As such, one might envision dark matter as the skeleton and backbone of space-time and dark energy as the force that is maintaining the integrity and stability of space-time skeleton.

In contradistinction to regular matter which is the condensed form of regular energy, dark matter and dark energy have no such relation.

They are born simultaneously out of the expansive force of big bang. They could be considered the two faces of the same coin and could represent quantum fluctuations of each other.

In other words, dark energy is entangled in the matrix of dark matter as dark vibrations and probably resides inside a wave function which comprises both.

Consequently, in the quantum universe, dark matter and dark energy could be buried in a quantum wave with peaks and troughs.

The peak and trough of this quantum wave would represent only one component and not both simultaneously, and in between there are different gradations of both.

In this sense, we would be able to measure only the wave function of either dark matter or dark energy along their quantum landscape fluctuation at any given point in time.

As such, one could potentially extend the general concept of Heisenberg's uncertainty principle, which was originally applied to speed and position of particles to this issue as well.

Thus, the more precise we measure one, the less precise we could measure the other one and vice a versa.

In other words, these two entities constantly flip-flop such that dark matter and dark energy constantly change identity.

This translates into a dark vibrating universe onto which regular energy and matter have been mounted.

The dark quantum wave representing dark energy and dark matter draws a sharp contrast with the regular virtual particles.

Such virtual particles are said to come into existence and get annihilated in the quantum universe continuously.

One might imagine that their coming into existence and annihilation is due to the wave function of dark matter and dark energy.

Thus, at the peak of the wave function where dark energy is dominant, virtual particle comes into existence and at the trough where dark matter is dominant, virtual particle gets annihilated.

The reason why it is critically important to understand such details is simply because the living cell is like a fish in the vast ocean of the known universe.

Our attempt at understanding the fish without a basic understanding of the ocean is a fallacy.

In turn, our attempt at understanding the sick fish, which is a disease state such as cancer, becomes even a bigger fallacy.

Currently, development of new cancer therapeutics ignores the fish and ocean concept.

In addition, it completely separates itself from the mechanisms that have led to the buildup of the known universe. This could be considered a fatal mistake.

Such strategies try to create new treatment modalities based on dissecting the details of sick fish as though there is no ocean.

The fundamental universal principles involved in creation and homeostasis of normal fish are often ignored in design of such treatment strategies as well.

This is one of the major reasons for the existence of insurmountable barriers between current cancer therapeutics and achievement of cure of cancer.

Cells of all living organisms and their constituents are made of nothing but vibrating superstrings.

Such vibrating superstrings are orchestrated in such a way to maintain maximum allowable free energy.

This takes place in an open thermodynamic system. Indeed, a living structure is defined as such.

We need to consider the basic concepts of universal laws that apply to creation of ocean or the known nonliving universe, and fish or the living universe, before jumping prematurely on the mission of understanding and curing cancer.

CHAPTER 1 ADDITIONAL READING

Afrasiabi, K., YiHong Zhou., and Angela Fleischman. 2016. GPCRs, at the crossroad of distortions in extracellular microenvironment and intracellular energetics homeostasis, a new model for twenty-first century cancer therapeutics. Cancer and Clinical Oncology 5, 31–42.

Atkins, Peter. 2007. *Four Laws - that Drive the Universe.* Oxford: Oxford University Press.

Berry, M. V., Klein, G., 1984. Newtonian trajectories and quantum waves in expanding force fields. J. Phys. A: Math. Gen. *17*, 1805.

Busch, P., Heinonen, T. and P. Lahti 2006. Heisenberg's Uncertainty Principle. 15th UK and European Meeting on the Foundations of Physics, Leeds, 29–31.

Fernflores, Francisco 2012. The Equivalence of Mass and Energy. The Stanford Encyclopedia of Philosophy. Edward N. Zalta (ed.)

Deupi, X., and B. K. Kobilka. 2010. Energy land-scapes as a tool to integrate GPCR structure, dynamics, and function. Physiology (Bethesda) *25*, 293–303.

Greene, B. 1999. *The Elegant Universe: Superstrings, Hidden Dimensions, and the Quest for the Ultimate Theory.* New York: Norton.

Greene, B., D. Morrison, and J. Polchinski. 1998. String Theory. Proceedings of the National Academy of Sciences of the United States of America 95.19, 11039–11040.

Hanahan, D., and R. A. Weinberg. 2011. Hallmarks of cancer: the next generation. Cell *144*, 646–674.

Hernandez, M., T. Ma, S. Wang. 2015. Theory of Dark Energy and Dark Matter. J. Math Study, Vol. 48, No. 3, pp. 199–221.

Ma, T. and S. Wang. 2014. Gravitational field equations and theory of dark matter and dark energy. Discrete and Continuous Dynamical Systems, Ser. A, 34(2): 335–366.

Macaulay, W. H. 1913. *The Laws of Thermodynamics.* Cambridge University Press.

McEvoy, E., V. S. Deshpande, and P. McGarry. 2017. Free energy analysis of cell spreading. Journal of the Mechanical Behavior of Biomedical Materials *74*, 283–295.

Vander Heiden, M. G., L. C. Cantley, and C. B. Thompson. 2009. Understanding the Warburg effect: the metabolic requirements of cell proliferation. Science *324*, 1029–1033.

Warburg, O. 1956. On respiratory impairment in cancer cells. Science *124*, 269–270.

CHAPTER 2

Entropy, Normal State, and Cancer

Second law of thermodynamics is considered the most fundamental law that governs the known universe at all levels of macro and micro cosmos.

Living cell has been recognized as the only machinery on the face of the known universe in which the speed of rise in entropy is the lowest.

Entropy could be defined as the degree of disorderliness, and it inversely correlates with the available free energy.

However, living cells surrender to a minimum rise in entropy. This threshold is set by the limits of this law.

The capability to minimize the increase in entropy could be considered the unique attribute of living cell.

The puzzle of birth of life and this exceptional capability seem to be intertwined. In other words, one cannot be separated from the other.

Indeed, any machinery capable of slowing the speed of rise in its entropy the most, develops the attributes of life and the one that is lively has already slowed down the pace of rise in entropy the most. Thus, living universes might represent an attempt at resisting the incessant rise in universal entropy.

Known living universes could not originate from dark matter, which comprises around 96 percent of the matter in known universe.

The reason is that dark matter is the static fabric or matrix of the known universe onto which regular matter keeps floating.

In contrast, the floating regular matter gets aligned with the thermodynamics arrow of time.

In other words, the second law of thermodynamics applies to regular matter and not to the vibrating dark matter which has filled up space-time.

As said before, all living cells and organisms are bound by the limits set by the second law. That is why they all get old, sick, and die.

Old age comes with significantly higher entropy as does cancer. That is why cancer is much more common in the elderly.

The main difference between an old cell and a cancer cell is the difference in their speed of rise in entropy. As such, cancer cell has aged much more rapidly.

In other words, time along the thermodynamics arrow of time has moved forward at tremendously higher speed in a cancer cell as compared with an old cell.

Consequently, the entropy of an old cell has increased gradually and over a much longer period and at much slower speed as compared with a cancer cell.

As such, aging results from gradual pile up of the minimum and unescapable amount of entropy, as dictated by the limits of the second law of thermodynamics.

Such increase in entropy could affect a diverse range of intracellular compartments and their function.

This could manifest itself as maladaptive ill mutations, as well as distorted quaternary structure of proteins with low free energy, which is associated with their dysfunction.

Such aberrancies and distortions could ultimately promote malignant transformation.

Thus, it seems plausible to conclude that to decrease the incidence of cancer, we need to slow down aging

It is also conceivable that as the aging cell piles up more and more entropy, at one point genomic instability would become inevitable.

By the same token if we could develop the capability to decrease the entropy of the cancer cell to the level of its normal counterpart, we might become able to convert a cancer cell to a normal cell.

One needs to keep in mind that the state of gene regulatory network is determined and governed by the dominant cellular energetics.

Thus, the increase in the free energy of the cellular networks, would translate into modification of the quaternary structure and physical orchestration of different sub compartments of the cell.

Consequently, we could potentially generate a shift of the whole genomic signature of the cell as well as its comprising networks toward normal state.

Clearly, this is a diffuse process which incorporates numerous intra cellular and transmembrane machineries ranging from microRNA to protein structure.

Physiological aging is associated with DNA damage and shortening of the telomeres. Cancer cell in contrast while having extensively damaged DNA, has long telomeres.

The reason for this dichotomy between cancer cell and old cell is the breakdown of normal inverse relationship between telomere length and aging.

Normal aging-related shortening of telomere length is a built-in defense mechanism against neo-plastic transformation.

Thus, at a certain critically short length of telomere, apoptotic machinery gets activated and the cell dies before age related mutations could lead to chromosomal instability and malignant transformation.

In cancer cell, accelerated aging or enormous increase in pace of entropy exceedingly supersedes normal pace of shortening of telomere.

Thus, apoptotic machinery cannot get activated and piled up mutations lead to chromosomal instability and neoplastic transformation.

This is achieved through convergence of two major events, namely inactivation of apoptotic machinery and increase in telomere length.

Mutation of P53 and activation of telomerase, are among prominent contributory factors.

Among the several known DNA damaging agents, oxidative damage is the most prevalent type.

Some of the other known mechanisms of DNA damage include alkylation, DNA intercalation, single and double strand DNA break and DNA mismatch.

However, they are not as dominant in the general population and not prevalent under normal condition.

The common denominator among all the DNA-damaging agents is that they lead to chromosomal instability, which is the hallmark of all cancers.

Chromosomal instability is the harbinger of carcinogenic mutations such as those that lead to activation and over expression of oncogenes and those that lead to dysfunction and loss of function of tumor suppressor genes.

In humans, generous physical activity and a diet rich in antioxidants are among the important factors that could potentially slow down DNA damage and aging.

This could potentially lead to a decrease in cancer risk. The protective effect of generous physical activity against aging and cancer seem to be multifaceted.

Most importantly, the very high oxygen consumption and consequent metabolomics signature seem to play a central role.

In other words, there would be a significant increase in the cellular network free energy.

This represents itself as a considerable increase in the number of available intracellular ATP molecules, which inversely correlates with cellular network entropy.

Hence, this can't be looked at as just a hypothesis, simply because even the resting basal metabolic rate of the population that is involved in generous physical activity is significantly higher than the control population.

The availability of more ATP molecules represents higher cellular network free energy at any given point in time.

This resembles a high voltage as compared with a low voltage running in a circuit

Consumption of substrates that could otherwise feed growth promoting pathways such as M-tor is among some of the positive contributory factors of generous and regular physical activity.

In addition, generous physical activity has potent anti-inflammatory effects.

Chronic inflammation is an important path leading to malignancy.

This path is also abrogated by generous and regular physical activity.

One could reasonably argue that the anti-inflammatory effects of regular generous physical exertion as well as its enhancement of oxidative phosphorylation would significantly decrease the number of free oxygen radicals and contribute to the increase in the cellular network free energy.

Some of the corollaries of increased intracellular network free energy, include suppression of oncogenic pathways, such as Ras and Wnt.

By the same token, tumor suppressor pathways such as Rb and p53 become uninhibited.

This is defined by what I have come to coin as Afrasiabi law of spontaneity.

This law dictates that under maximum allowable and achievable intracellular free energy, all cellular and molecular sub-compartments get orchestrated accordingly to maintain that state.

In other words, molecular evolution has led to fascinating selection of molecular components of the cellular network, which extends from but is not limited to genome, epigenome, spliceosomes, and microRNA network.

The selected components communicate elegantly at different levels of cellular network free energy to achieve the final common goal of maintenance of maximum allowable free energy.

Such harmony, balance, and orchestration could be modulated through changes in the physical cues in subcellular and molecular compartments, which govern the dynamics of their relationship and communication.

As such, tumor suppressor genes and proteins, as well as oncogenes and oncoproteins behave accordingly.

This could be considered the big miracle of living universes. In other words, all the critical pieces of the biological machines are designed and programmed accordingly.

As one example, the spliceosomes that determine the makeup of different mRNAs, function differently under different intracellular network free energy levels to generate proteins with maximum allowable free energy.

The hyperplasticity of the living cell can generate the opportunity to modulate the same function at different subcellular levels from epigenome all the way to microRNA network.

This hyperplasticity generates numerous collateral pathways, so that if one could not achieve the goal, the other would.

Thus, Rb and p53 pathways, as well as Ras and Wnt pathways among many others get employed toward maintenance of homeostasis of cellular energetics under normal condition.

In other words, the governing rule satisfies the final goal of maintenance of the highest possible cellular network free energy which inversely correlates with cellular network entropy.

Cancer could be considered the breakdown of this harmony. Such breakdown would not allow achievement of the highest possible cellular network free energy.

The rapid pace of cancer cell proliferation is a futile attempt at minimizing cellular network entropy.

The randomness of gene expression, with its resultant protein-protein interactions contributing to local network entropy has been shown to be higher in cancer cell as compared with normal cell.

Genes which drive cell-proliferation in cancer cells and which often encode oncogenes are associated with reductions in network entropy.

The strategies currently employed in treating cancer, ignore the deep disturbance in cancer cell network entropy and its inherent resistance to death.

This prevailing philosophy dictates that killing cancer cells would lead to elimination of tumor mass and ultimately cure.

This concept has faced major barriers. This philosophy has several other fundamental flaws as well.

First and most importantly, it ignores the intratumoral heterogeneity, which enables some cancer cells to escape this attempt.

Secondly, it ignores the hyperplasticity of cancer cell which originates from its inherent chromosomal instability.

Thirdly, it ignores the deeper concept that cancer cell has evaded all the barriers and has escaped to infinity, while cytocidal modalities are simply finite measures.

In other words, cancer cell has already reversed the evolutionary path of living universes that is founded on maintaining the cellular network entropy at the lowest possible level and is marching successfully in the opposite direction.

CHAPTER 2 ADDITIONAL READING

Afrasiabi, K., Y. H. Zhou, and A. Fleischman. 2015. Chronic inflammation: is it the driver or is it paving the road for malignant transformation? Genes Cancer 6, 214–219.

Clark-Matott, J., A. Saleem, Y. Dai, Y. Shurubor, X. Ma, A. Safdar, M. F. Beal, M. Tarnopolsky, and D. K. Simon. 2015. Metabolomic analysis of exercise effects in the POLG mitochondrial DNA mutator mouse brain. Neurobiol Aging 36, 2,972–2,983.

Davies P., E. Rieper, J. Tuszynski. 2013. Self-organization and entropy reducing in a living cell. Biosystems 111, 1–10.

De Lange, T. 1994. Activation of telomerase in a human tumor. Proc Natl Acad Sci U S A *91*, 2,882–2,885.

Dell'Orco D., D. Casciari, F. Fanelli. 2008. Quaternary structure predictions and estimation of mutational effects on the free energy of dimerization of the OMPLA protein. J. Struct. Biol. *163*,155–162.

Garrett, K. and G. Duda. 2011. Dark matter: a primer. Advances in Astronomy. 968283

Giardini, M. A., M. Segatto, M. S. da Silva, V. S. Nunes, and M. I. Cano. 2014. Telomere and telomerase biology. Prog Mol Biol Transl Sci *125*, 1–40.

Hanselmann R. G., and C. Welter. 2016. Origin of Cancer: An Information, Energy, and Matter Disease. Front Cell Dev Biol. *4*, 121.

Holland, A. J., and D. W. Cleveland. 2009. Boveri revisited: chromosomal instability, aneuploidy and tumorigenesis. Nat Rev Mol Cell Biol *10*, 478–487.

Nasi, M., A. Cristani, M. Pinti, I. Lamberti, L. Gibellini, S. De Biasi, A. Guazzaloca, T. Trenti, and A. Cossarizza. 2016. Decreased Circulating mtDNA Levels in Professional Male Volleyball Players. Int J Sports Physiol Perform 11, 116–121.

Rajagopalan, H., and C. Lengauer. 2004. Aneuploidy and cancer. Nature *432*, 338–341.

Rhyu, M. S. 1995. Telomeres, telomerase, and immortality. J Natl Cancer Inst *87*, 884–894.

Roake, C. M., and S. E. Artandi. 2017. Control of Cellular Aging, Tissue Function, and Cancer by p53 Downstream of Telomeres. Cold Spring Harb Perspect Med 7.

Soares, J. P., A. M. Silva, M. M. Oliveira, F. Peixoto, I. Gaivao, and M. P. Mota. 2015. Effects of combined physical exercise training on DNA damage and repair capacity: role of oxidative stress changes. Age (Dordr) 37, 9,799.

Tarabichi, M., A. Antoniou, M. Saiselet, J. M. Pita, G. Andry, J. E. Dumont, V. Detours, and C. Maenhaut. 2013. Systems biology of cancer: entropy, disorder, and selection-driven evolution to independence, invasion and "swarm intelligence." Cancer Metastasis Rev. 32(3–4), 403–421.

West, J., G. Bianconi, S. Severini, and A. E. Teschendorff. 2012. Differential network entropy reveals cancer system hallmarks. Sci Rep 2, 802.

Xu, L., S. Li, and B. A Stohr. 2013. The role of telomere biology in cancer. Annu Rev Pathol *8*, 49–78.

CHAPTER 3

Aneuploidy, Microenvironment, and Cancer

In evolutionary biology, solid tumor at its advanced stage is best exemplified as a society, a society that each of its members could repopulate the whole heterogeneous tumor population. This is one of the most puzzling and unique features of cancer.

This unique evolutionary feature of tumor mass is based on its exceptional capacity to transfer and acquire the necessary survival factors from its unfit and dying cells to the surviving neighboring cells.

Survival factors are mostly transferred by genetic material coding for such survival factors.

Some relevant examples include double minute carrying EGFR in glioblastoma multiforme (GBM), and parts of and on occasion the whole chromosome in other neoplastic disorders such as leukemia.

As such, aneuploidy, which is a common and nonrandom feature of some of the neoplastic disorders represents one major mechanism.

Aneuploidy is one of the hallmarks of chromosomal instability, the common denominator of neoplastic disorders.

In contradistinction to the acquired aneuploidy, congenital aneuploidy in disorders such as Down's syndrome does not correlate with chromosomal instability.

The significantly elevated risk of malignant transformation in such patients is related to higher gene dose, and its ensuing aberrancies.

The cancer-specific chromosomal instability gives rise to clonal and non-clonal chromosome aberrations.

As such, it fuels the speed of cancer evolution, maintains tumor heterogeneity, and supports the generally observed cancer resistance to targeted and cytotoxic therapy.

It is conceivable that the aging cell also piles up more and more cellular network entropy, and at one point chromosome instability would ensue.

Chromosomal instability, is the harbinger to further increase in cellular network entropy.

The increase in cellular network entropy affects gene expression machinery accordingly. Aneuploidy could also arise under such conditions.

The resulting ill humeral factors spill into the adjacent microenvironment and lead to further complication of underlying pathology.

Consequently, microenvironment also undergoes drastic changes through both biochemical as well as the resultant physical cues.

These sequence of events signify the inseparable nature of relationship between the malignant cell and microenvironment.

Hence, these two compartments relate with each other the way fish and ocean do.

To further clarify this concept, one relevant example is aneuploidy of chromosome 7 in GBM, which incidentally correlates with aggressive biological behavior of this malignancy.

In this instance, the pathological interaction between microenvironment and cancer cell is modulated through extracellular matrix proteins (EMPs) such as fibulins.

The foundation of this pathological interaction is aneuploidy induced breakdown of the balanced and fine-tuned energetics relationship of normal cell and microenvironment.

Hence, the availability of extra genetic material because of chromosome 7 aneuploidy, which codes for extra copies of EGFR seems to play a central role in this regard.

This breakdown is the dominating force governing the pathological relationship between cancer cell and its microenvironment.

Because of this breakdown, tumor microenvironment also becomes highly disorganized.

In case of GBM, this disorganization is the result of interaction of fibulins, such as EFEMP1 with other microenvironment matrix proteins.

This is preceded by the dysregulated expression of EMPs such as EFEMP1 (EGF containing fibulin like extracellular matrix protein 1), which is one of the many matrix proteins.

Incidentally, EFEMP1 dysregulation has been reported in around 40 percent of solid malignancies.

In case of GBM, this dysregulation seems to be preceded by overexpression of EGFR due to underlying chromosome 7 aneuploidy.

In this case, aneuploidy secures the survival of cells neighboring the dying cells with high network entropy by securing downstream survival signals through overexpression of EGFR.

To sum up the sequence of events:

1. Cellular network entropy irreversibly increases because of a major mishap.

2. Chromosome 7 or double minutes which contain EGFR are released by the dying cells with high network entropy and acquired by neighboring cells, to secure survival of those cells.

3. Fibulins such as EFEMP1 get overexpressed in the surviving cells, in response to overexpression of EGFR to secure the downstream survival signals in such cells.

4. Fibulins (e.g., EFEMP1, FBLN5) get released into the microenvironment and together with their regulated MMPs (matrix metalloproteinases) drastically alter the physical cues of the microenvironment toward a much more disorderly state.

This is mostly based on the unique and specific feature and function of fibulins, namely intramolecular bridging of the microenvironment matrix proteins and cell membrane receptors.

This is expected to ultimately lead to modification of the energetics of the microenvironment, in favor of a significant decrease of its free energy.

One potential explanation for decrease in microenvironment free energy is alteration of elasticity and orderliness of microenvironment by distorted intramolecular bridging of matrix proteins.

As such, the disorderly microenvironment, would offer much less opportunity for plasticity as demanded by constant changes in the microenvironmental physical cues.

This type of microenvironment could be looked at as one with low free energy and high entropy, represented by lesser degree of plasticity and significantly higher disorderliness.

This would get further complicated by the digestive function of MMPs, which ultimately pave the way for invasion of cancer cells into surrounding tissue and distant organ structures.

Hence, the road gets paved for cells that cannot grow and thus need to go.

Consequently, aneuploidy would enable the neoplastic cell to survive and metastasize to surrounding structures and tissues.

Because of the above-mentioned sequence of events, the increase in intracellular network entropy spreads like a wave to the adjacent microenvironment.

In summary, aneuploidy not only secures survival of the malignant cells but in addition it is the guardian of chromosomal instability.

In addition, it promotes the spread of increased cellular network entropy to the microenvironment and facilitates metastatic spread of the malignant cell.

On top of the physical modifications enumerated above, fibulins, such as EFEMP1, also interact with membrane receptors, such as EGFR, IGF1R, and NOTCH.

These interactions would activate downstream pathways such as PI3Kinase, which would ultimately contribute to proliferative and invasive features of cancer cells.

Convergence of modulation of physical cues of the microenvironment and biological modification of cellular pathways by EFEMP1 is quite fascinating.

This convergence empowers EFEMP1 with unique attributes and potentially explains why in around 40 percent of solid malignancies there is dysregulation of this protein, which is only one of the many matrix proteins.

In addition, reversal of such effects by EFEMP1-derived tumor suppressor protein (ETSP), which becomes dysfunctional in malignant cell is even more intriguing.

This example denotes that the breakdown of Afrasiabi law of spontaneity, which I alluded to before, would put the cell at maximum disadvantage.

In other words, by convergence of the released antagonizing genes, molecules and pathways, the direction of spontaneity law gets reversed.

Consequently, cooperation among the released antagonizing forces drive the cell toward malignant transformation.

Procurement of the highly invasive and proliferative nature of GBM by a permissible and distorted microenvironment with high entropy is one of the many examples in neoplastic disorders.

Thus, it appears that EFEMP1 has acquired the unique evolutionary advantage and capability of goal oriented alteration of physical and biochemical cues simultaneously.

Alternatively, one might say that under the selection pressure of the very high cellular network entropy of GBM cells, EFEMP1 has emerged to promote these functions.

Even more fascinating, EFEMP1 has context-dependent dual functions in cancer.

As such, through mechanisms such as alternate splicing, it could play both the role of oncogene and tumor suppressor gene as dictated by the context.

It is most plausible that context would be represented by the cellular network energetics, which is the foundation of cellular matrix.

Consequently, under high cellular network free energy, the splice variant that dictates high proliferation rate would get suppressed.

By the same token, under low cellular network free energy, the splice variant that promotes low proliferation rate would get suppressed.

This variation or dual function also simultaneously respects the homeostasis of intratumoral heterogeneity such that the tumor mass forming cells and the stem cell-like cells would have different variants satisfying dormancy of stem cell-like cells and high proliferation rate of tumor mass like cells.

Theoretically, such evolutionary selection pressure affecting fibulins might exert its role at different levels ranging from alternate splicing of fibulin MRNA to specific modification of epigenome or microRNA network toward achievement of the same goal.

This concept, which might be more appealing, gives the high neoplastic cellular network entropy the primary or driving role.

Under such premises, the micro environmental modifications acquire a secondary role guided by the primary driving force of high neoplastic cellular network entropy.

This concept, could play a game changing role as far as our future design of cancer therapeutics is concerned.

In other words, we would go after modification of malignant cell network entropy.

Under these premises, directly targeting the key micro environmental player molecules such as EMPs shaping the malignant microenvironment might not be an attractive primary or pertinent goal.

Alternatively, primary derangement of the microenvironment leading to high microenvironment entropy could potentially convey similar cues to the cellular network entropy.

In this case ill events originating in the micro environment and increasing its entropy could act as the driving malignant force and become a pertinent target for cancer therapeutics.

Theoretically, events leading to malignant transformation could potentially start in either the microenvironment or the cell destined to become malignant and even their combination.

By the token of the above premises, future cancer treatment strategy should be based on better understanding and elucidation of the compartment of initiation of neoplastic transformation in each specific malignancy.

Recent discovery of the role of EFEMP1 in differentially regulating respiration profile of different cells comprising the GBM tumor mass, lends support to the critical role of fine tuning of cellular energetics as the main mechanism of maintaining intra tumoral heterogeneity.

Either the distorted energetics of microenvironment or cell as the driving force of malignant transformation, could present the same way and lead to the same outcome.

It is possible that our thinking about assigning a primary site of initiation or triggering event of carcinogenesis could be elusive.

In other words, the microenvironment and cell are so closely intertwined and entangled that a major distortion in one would get transferred to the other one instantaneously.

In this regard, dynamics of modifications and interactions of EFEMP1 within the cell and with the microenvironment respectively, might offer a good example.

In either scenario, oxidative phosphorylation is upregulated in tumor mass forming cells and aerobic glycolysis which is represented by Warburg effect is activated in tumor initiating cells through differential expression of EGFR and NOTCH by EFEMP1 in such cells.

Thus, NOTCH expression is upregulated in stem-like tumor initiating cells while EGFR is upregulated in tumor mass forming cells.

This difference in EFEMP1 mediated overexpression of EGFR and NOTCH is determined by the inherent difference in energetics bar code of these two cell populations.

As such, the distorted energetics of GBM microenvironment could affect and modify the energetics of different cells comprising the GBM tumor mass accordingly and vice versa.

One could also argue that abnormal quantity and/or splice variants of EMPs are made by the neoplastic or transformed cells and released into the micro environment.

In this case, as mentioned before the cell and not the micro environment could be considered the driving force and the starting point.

In other words, the transformed cell as the driving force would lead to the neoplastic changes of the micro environment through mediators such as EMPs.

By the same token, as mentioned above the distorted and driving micro environment force could also dictate neoplastic and transforming changes of the cell.

This could happen by the interplay between the preexisting different cellular network entropy bar codes and a new high entropy microenvironment.

This difference in cellular network entropy bar codes is generated by difference in the physical location and physical cues of different cells comprising the tumor mass.

The variation in these energetics bar codes would lead to generation of different splice variants of EMPs as well as different genetic signatures in different cells comprising the tumor mass.

This is probably a major determinant of intra-tumoral heterogeneity and innumerable genetic signatures of the cells comprising the tumor mass.

Either the transformed cell or the transformed microenvironment would modify the other compartment accordingly to sustain and perpetuate progression of the neoplastic transformation.

This represents a goal-oriented and orchestrated collaboration between these two inseparable compartments.

As said before, GBM could serve as a good model for elucidation of the potential primary driving force and dynamics of interaction between the cancer cell and microenvironment.

In this regard, the simple lead into intratumoral heterogeneity of GBM is best represented by the extent of chromosome 7 aneuploidy.

This is representative of the degree of chromosomal instability and is a manifestation of the distorted energetics and unique physical features of GBM tumor cells and their microenvironment.

Consequently, one of the major prognostic factors in astrocytomas is the degree and extent of chromosome 7 aneuploidy.

Prognosis is clearly worse in GBM because it carries a higher degree of chromosome 7 aneuploidy as compared with low and intermediate grade gliomas and anaplastic astrocytoma.

The role played by physical cues and the ensuing intratumoral heterogeneity represented by innumerable genetic imprints of the cells comprising the tumor mass is a fascinating issue and deserves special attention.

In this sense, the position of the tumor cells in the tumor mass, which translates into different physical cues such as angulation, pressure, or tension is of critical significance.

As such, this could ultimately translate into variable gradations of cellular network entropy of different tumor cells which comprise the tumor mass.

This variation in cellular network entropy translates into variable gradations of chromosomal instability

As said earlier, in case of GBM, it is best represented by the degree of chromosome 7 aneuploidy in different cells comprising the GBM tumor mass.

The relationship between the physical cues of individual cells comprising the tumor mass and their genetic signature and degree of Chromosomal instability, also has strong bearing on repopulation of the whole tumor mass by a few surviving cells following chemo, radio, or immunotherapy.

Hence, repopulation of the whole tumor mass with the same distribution and degree of variation in genetic signatures is the immediate result of difference in physical cues of off springs of the few repopulating cells in tumor mass.

This happens simply because no two daughter cells could occupy the same physical location in the tumor mass.

Thus, each cell would have a different physical cue or entropy bar code, leading to and manifested by a different genetic signature.

In case of astrocytomas, different gradations of chromosome 7 aneuploidy represent such variations.

Thus, repopulation of tumor mass does not necessitate a specific preexisting genetic programming of the few surviving cells to enable them to repopulate the tumor mass with cells having similar variations in genetic signatures.

In addition, variations in expression of a specific EMP (such as EFEMP1) guided by variations in physical cues of those cells could direct and reciprocate the tumor microenvironment's driving force on tumor mass accordingly.

This interplay between the tumor mass and microenvironment is so intimate that at times it is hard to distinguish and see them as separate compartments.

One could conclude that aneuploidy is a major and common representation of breakdown of normal relationship between these two compartments.

Consequently, aneuploidy gives birth to one of the common paths that leads to the up rise of malignant transformation.

CHAPTER 3 ADDITIONAL READING

Duesberg, P., D. Mandrioli, A. McCormack, and J. M. Nicholson. 2011. Is carcinogenesis a form of speciation? Cell Cycle *10*, 2,100–2,114.

Duesberg, P., R. Stindl, and R. Hehlmann. 2001. Origin of multidrug resistance in cells with and without multidrug resistance genes: chromosome reassortments catalyzed by aneuploidy. Proc Natl Acad Sci USA *98*, 11,283–11,288.

Gryder, B. E., C. W. Nelson, and S. S Shepard, 2012. Biosemiotic Entropy of the Genome: Mutations and Epigenetic Imbalances Resulting in Cancer. Entropy 15, 234–261.

Heng, H. H., S. W. Bremer, J. B Stevens, S. D. Horne, G. Liu, B. Y. Abdallah, K. J. Ye, and

C. J. Ye. 2013. Chromosomal instability (CIN): what it is and why it is crucial to cancer evolution. Cancer Metastasis Rev *32*, 325–340.

Hu, Y., C. Ke, N. Ru, Y. Chen, L. Yu, E. Siegel, M. Linskey, P. Wang, and Y. Zhou. 2015. Cell context-dependent dual effects of EFEMP1 stabilizes subpopulation equilibrium in responding to changes of in vivo growth environment. Oncotarget *6(31)*, 30,762–30,772.

Hu, Y., P. D. Pioli, E. Siegel, Q. Zhang, J. Nelson, A. Chaturbedi, M. Mathews, D. Ro, S. Alkafeef, H. Nelson, M. Hamamura, L. Yu, K. Hess, B. Tromberg, M. Linskey, and Y. Zhou. 2011. EFEMP1 suppresses malignant glioma growth and exerts its action within the tumor extracellular compartment. Mol Cancer 10, 123.

Hu, Y., N. Ru, H. Xiao, A. Chaturbedi, N. T. Hoa, X. J. Tian, H. Zhang, C. Ke, F. Yan, J. Nelson, et al. 2013. Tumor-specific chromosome mis-segregation controls cancer plasticity by maintaining tumor heterogeneity. PLoS One 8, e80898.

Kops, G. J., B. A. Weaver, and D. W. Cleveland. 2005. On the road to cancer: aneuploidy and the mitotic checkpoint. Nat Rev Cancer 5, 773–785.

Nicholson, J. M., and D. Cimini. 2011. How mitotic errors contribute to karyotypic diversity in cancer. Adv Cancer Res 112, 43–75.

Roth, J. J., T. M. Fierst, A. J Waanders, L. Yimei, J. A. Biegel, and M. Santi. 2016. Whole Chromosome 7 Gain Predicts Higher Risk of Recurrence in Pediatric Pilocytic Astrocytomas

Independently From KIAA1549-BRAF Fusion Status. J Neuropathol Exp Neurol *75(4)*, 306–615.

Tsiambas, E., N. Mastronikolis, A. Lefas, N. S. Georgiannos, V. Ragos, P. P. Fotiades, N. Tsoukalas, N. Kavantzas, A. Karameris, D. Peschos, E. Patsouris, and K. Syrigos. 2017. Chromosome 7 Multiplication in EGFR-positive Lung Carcinomas Based on Tissue Microarray Analysis. In Vivo. *31(4)*, 641–648.

Valind, A., Y. Jin, B. Baldetorp, and D. Gisselsson. 2013. Whole chromosome gain does not in itself confer cancer-like chromosomal instability. Proc Natl Acad Sci USA *110*, 21,119–21,123.

Zhou, Y., Y. Hu, L. Yu, C. Ke, C. Vo, H. Hsu, Z. Li, A. T. Di Donato, A. Chaturbedi, J. W. Hwang,

E. R. Siegel, and M. Linskey. 2016. Weaponizing human EGF-containing fibulin-like extracellular matrix protein 1 (EFEMP1) for twenty-first century cancer therapeutics. Oncoscience, 3(7-8), 208–219.

CHAPTER 4

Location, the Seed of Function, and Fate Specification

Location of cell in tissue and organoid at all levels and time frames since the beginning of embryogenesis carries a critical significance.

Physical cues such as torque, angulation, and tension as well as proximity to niche cells are all determined by location.

There is clear-cut evidence that such physical cues directly affect gene regulatory mechanisms at all levels, ranging from epigenome to microRNA and spliceosomes.

Proximity to niche cells facilitates transfer of growth and survival promoting factors. In addition, location plays a central role in conveyance of micro-environment cues to the cell.

Deep understanding, and application of such fundamental concepts and principles governing the homeostasis of normal state, could open the door on a new chapter in cancer therapeutics.

This would ultimately change our way of thinking about all aspects of tumor biology. It would also affect our daily simple definitions, which we have taken for granted.

In this regard, one notorious example is the current definition of stem cell and its identity.

Currently, somatic stem cell is defined as a cell with self-renewal capability. In addition, as dictated by variations and circumstances, its other attribute of asymmetrical division could also become manifest.

Current definition of stem cell seems to ignore that the location or position of that specific cell in tissue is the main factor and the driving force that creates its unique features of stemness.

Indeed, there are no specific features that are unique to that cell, which could not get recapitulated by another cell put in the same location and specific condition.

As one example, the few cells at the base of the crypt of the large bowel, which is composed of around three thousand cells, have stemness capability.

These cells are in close contact with myofibroblasts which convey the unique features, humeral factors and cues of microenvironment to those few cells.

Numerous humeral mediators which might include FGF and EGF and perhaps EMP's contribute to transformation of those few cells into crypt stem cells.

These few cells which have renewal capability are sitting next to Paneth cells. They could repopulate the whole crypt and it seems that neoplastic transformation happens in those cells or the cells that transform to crypt stem-like cells.

Experiments done through high throughput technology following destruction of malignant crypt stem cells (so-called cancer stem cells) have failed to prevent the repopulation of the malignant crypt by the surviving non-stem cells.

Such experiments have surprisingly demonstrated that surviving cells such as a transit amplifying cancer cells that are on the top portion of the crypt can turn into cancer stem cells and repopulate the whole malignant crypt.

There are at least two potential explanations for this astonishing observation:

1. Relocation of surviving cells such as transit amplifying cancer cells (TACS) to the bottom of crypt and acquiring stemness capability by getting into close contact with niche cells and the necessary micro environmental cues.

2. Another possibility is the transfer of humeral mediators of dying stem cells to cells such as TACS and, hence, changing their location and/or identity to that of a stem cell.

Regardless of such details, location of the cell seems to play a critical role as far as its identity is concerned.

As mentioned before, in evolutionary biology, tumor is akin a society, a society that is programmed on growth and survival.

This model, is a universal model and fundamental concept for homeostasis of living universes, also holds true for maintenance of integrity of any organoid or tissue.

Regeneration of stem cell either via transfer of humeral factors released at the time of demise of the stem cell or relocation of other cells to the unique physical position that endows upon them stem cell capability or a combination thereof, secures this principle and achieves this goal.

This is the foundation of the central concept of tumor as a society, a society in which any member can get reprogrammed to replenish the whole tumor mass.

A normal cell population, put under stress, is also expected to behave the same way. That is to say, the surviving normal nonstem cells follow the same principles and replenish the lost cells.

In this regard, as mentioned before physical location of the repopulated cells plays a critical role.

As such, cells positioned at different locations would receive different physical cues from microenvironment.

In addition, the physical cues related and specific to different positions would translate into different cellular network entropy bar codes.

The amount of cellular network entropy affects gene expression machinery of the cell accordingly.

Thus, cells acquire different genetic, epigenetic, and microRNA network profiles and identities and behave differently as delineated above.

The reciprocity between gene regulatory mechanisms and cellular network entropy, guarantees maintenance of the fine balance of cellular network energetics.

Variation in surrounding microenvionment among cells happens spontaneously simply because each offspring cell is occupying a different physical location and is receiving different physical cues. This would translate into innumerable genetic signatures in cells comprising normal tissue and tumor mass.

Another presentation of this principle is the hyperplasticity of tumor.

This originates from such variations in energetics and network entropy as well as genetic signatures of its comprising cells as dictated by different physical positions and microenvironmental cues conveyed on those cells inside the tumor mass.

One good example is differentiation of the bowel crypt stem cells located at the base of crypt into different cells such as goblet cells, Paneth cells, and transit amplifying cells lining the gastrointestinal tract.

Another puzzling role played by location is best exemplified by transformation of hematopoietic stem cells into cardiac myocytes following experimental implantation of such cells into the interstitium of heart.

Yet another example is proof of hair follicle stem cell acting as stem cell of acute lymphocytic leukemia in a well-documented case.

These examples offer solid evidence that physical location of the cell not only is associated with specific physical cues but would also lead to modification of genetic and biochemical machinery of the cell.

By the same token, differentiation of cells into ectoderm, mesoderm, endoderm, and their derived organs during embryogenesis has to do with the fine role that location and movement of cells play.

As such, there are broad ramifications of this concept, both as far as understanding of normal and abnormal states are concerned. In other words, location plays a central role in identity of cells as well as the organ comprised of those cells and their homeostasis.

Application of these principles could guide us in the design of future cancer and non-cancer therapeutics.

As one example, if we could implant the healthy stem cells of one organ into the stem cell niche of a diseased one, while suppressing the ill microenvironmental cues of the diseased organ, we might become able to treat a wide range of malignant disorders.

Potentially autoimmune and neurodegenerative disorders such as amyotrophic lateral sclerosis and Parkinson's disease, among many others, could be approached accordingly.

Current cancer therapeutic strategies is based on elimination of cancer cells through chemo, radio, and targeted therapy aiming at killing cancer cells.

Immunotherapy is based on activating the immune system to eliminate cancer cells. In this case the reason for ultimate failure and recurrence of malignancy is mostly in the failure of immune system to recognize cancer stem cells that are antigenically very similar to normal stem cells.

Tumor recurrence will be inevitable when tumor microenvironment is reestablished by surviving tumor cells capable of occupying a critical location.

Again, in this case if we could block the availability of the critical location to the surviving tumor cells, we might become able to abort recurrence of malignant disorder.

As will be discussed in future chapters, a very promising way to conquer cancer is conversion of malignant cells and their microenvironments to its normal counterpart.

This demands a much deeper understanding of the dynamics of normal state and interplay of normal and malignant cell with microenvironment, as well as mastering the details of location and conversion technology.

As such, at the root of these principles, there sits the critical role played by location of the cell in either normal tissue or tumor.

CHAPTER 4 ADDITIONAL READING

Angelini, P., and R.R Markwald. 2005. Stem Cell Treatment of the Heart: A Review of Its Current Status on the Brink of Clinical Experimentation. Texas Heart Institute Journal / from the Texas Heart Institute of St. Luke's Episcopal Hospital, Texas Children's Hospital, 32(4), 479–488.

Barker, N., R. A. Ridgway, J. H. Van Es, M. Van De Wetering, H. Begthel, M. Van Den Born, E. Danenberg, A. R. Clarke, O. M. Sansom, and H. Clever. 2009. Crypt stem cells as the cells-of-origin of intestinal cancer. Nature 457, 608–611.

Medema, J. P. and L. Vermeulen. 2011. Micro-environmental regulation of stem cells in intestinal homeostasis and cancer. Nature *474*, 318–326.

Rajasekhar, V. K., and M. C. Vemuri. 2009. *Regulatory Networks in Stem Cells.* Stem Cell Biology and Regenerative Medicine.

CHAPTER 5

Breakdown of Logic and
Birth of Catastrophe

The birth and evolution of living and nonliving universes are based on a handful of laws that govern the homeostasis of these universes.

The most fundamental law that governs the known universe at all levels, from micro to macro cosmos is the second law of thermodynamics.

As was alluded to in previous chapters, this law dictates that the degree of disorderliness increases along the thermodynamics arrow of time.

Living universes have the unique attribute of minimizing the pace of rise in entropy to the limits set by this law.

Alignment with these laws, and most specifically the second law of thermodynamics, is the foundation of what we perceive and define as the logical behavior of living and nonliving universes.

Any obviation from these laws, would lead to catastrophic conditions including disease states and premature demise, as well as cataclysmic events in the known nonliving universe.

As such, cancer is the breakdown of the logic of normal cell division rather than the illogical cancer cell division as an abstract phenomenon.

Consequently, our biggest handicap at decoding the puzzle of cancer is lack of deep understanding of the logic of normal cell division.

This in part originates from our abstract look at the cell and ignoring universal principles such as the fish and ocean concept.

This concept implies that all living organisms are like fish in the vast ocean of the known universe and understanding the fish demands deep understanding of the ocean.

Over the last several decades, we have gathered vast amount of detailed information about intracellular communication network, as well as different machineries, pathways, and molecules involved in mitosis.

The spectrum of our knowledge extends from cyclin-dependent kinases and their positive and negative control loops to mitotic spindle apparatus, Aurora kinases, trans-membrane kinases, TP53 and

BCL2 family, and telomere biology, to have mentioned just a few.

All these are components and control loops of the mitotic machinery of the cell.

Their main mission is homeostasis of cell division, in such a way that inappropriate expansion and shrinkage of the cell population is prevented.

Mitosis, while acting as the main machinery to generate offspring cells, also is an ingenious engine that maintains the maximum allowable free energy which correlates with the lowest entropy.

In contrast, during the same period, entropy of the nonliving universe, which includes the surrounding environment of the living cell, continues to increase at a much faster pace.

Thus, mitosis is the antientropy machinery of the living universes.

In other words, the distinguishing feature between the living and nonliving universes is their opposite alignments along the thermodynamics arrow of time.

In this sense, the nonliving universes are entropy-generating machineries while living universes are antientropy engines.

Other than this distinguishing feature, the astonishing similarities between these two universes could be seen at each point in their evolution.

The volcanic eruption of mitosis in living universes mostly starts following fertilization of ovum by sperm.

However unicellular organisms also continue with incessant replication in the absence of fertilization.

The driving force for cell division in both instances seems to be the same principle of keeping the cellular network entropy at minimum allowable level as per the limits of the second law.

In case of unicellular organisms, it is the incessant penetration of pro-entropy environmental forces through the barrier of cell membrane or cell wall that sets the pace of cell division.

In case of fertilized cell, it is the abrupt increase in cellular network entropy following fertilization that sets the pace and direction of mitosis.

In case of volcanic eruptions of mitosis in somatic stem cell compartments, it is a predetermined and preprogrammed chaos in this compartment that acts as the triggering mechanism.

In contrast, healthy asymmetrical somatic stem cell division leads to a decrease in cellular network entropy by its diffusion to a diversified cell population.

This acts as the foundation of homeostasis of organoids and organs. New generation of cells replaces the old ones and differentiation of progenitor cells secures the balance among different subpopulations.

Again, we could see interplay of chaos and harmony for maintenance of homeostasis in the quantum living and nonliving universes.

On the contrary, in the known nonliving universe, the main engine for proliferation or in other words generation of space-time is the entropy generating force of big bang, in contrast to the antientropy force of mitosis for generation of new living universes.

The stability of space-time and the birth of galaxies originate from entanglement of dark matter and dark energy.

Dark matter and dark energy constitute the matrix of the known universe and are the immediate derivatives of the force of big bang.

By the same token, the apparent stability of the fertilized cell originates from entanglement of ovum and sperm DNA, which generate the matrix of the fertilized ovum.

Again, in both living and nonliving universes, chaos promotes a transitional stable state.

The perceived stability of this transitional state is observer dependent and is the result of the way the observer looks at that state at any given point in time.

The observer itself is a chaotic entity made of innumerable short living stable cross sections.

In other words, a snapshot of that state and not the whole state or event is perceived by observer and analyzed.

That state is interpreted as stable because the observer is transiently stable enough to be able to look at that state and take a snapshot.

This is another attestation to the fact that in quantum living and nonliving universes, chaos and stability are the two faces of the same coin.

In other words, what we perceive as stable systems have their root deep in chaos, and chaos itself has innumerable stable cross-sections.

Deep understanding of this concept is of essence, as far as dissection and analysis of ill and healthy living and nonliving universes are concerned.

Cancer as a major proentropy event in antientropy living universes, and anti-entropy events such as black holes and neutron stars in proentropy nonliving universes are among some of the ill examples.

One needs to keep in mind that on both occasions, the ill events eventually get terminated.

In case of cancer, termination happens through demise of the host.

The resulting disintegrated and decomposed building blocks of living universes could potentially get recycled and used for the buildup of another healthy living universe.

In case of black hole, termination process starts through the entropy zone or event horizon of black hole.

Gradual evaporation and final elimination of black hole happens through escape of virtual particles with positive energy and mass generated near event horizon.

In contrast, entrapment of generated virtual particles with negative energy and mass adjacent to event horizon into the black hole is immediately responsible for loss of mass of black hole.

These concepts act as the foundation and language of the logic of both living and nonliving universes.

Living universes strongly resist the shift toward a heavier burden of chaos.

In contradistinction, the known nonliving universe aligns itself with heavier weight of chaos.

In living universes, the chaotic process of fertilization creates a foundation which triggers incessant rounds of mitosis.

This antientropy engine leads to generation of new living universes comprised of organs, akin galaxies of known nonliving universe.

Thus, the universal event that best defines fertilization is sudden and massive increase in chaos or entropy of fertilized ovum.

This originates in part from crowding in the space occupied by ovum. The population of chromosomes in human egg increases from twenty-three to forty-six instantaneously.

In addition, paternal genetic imprinting is totally different from maternal imprinting pattern.

This would act as an additional chaotic or proentropy force pushing for recurrent rounds of mitosis.

In other words, instantaneous physical crowding and biological diversity generated because of fertilization act as synergistic elements of chaos.

Thus, the increase in disorderliness is represented both biologically and physically in the closed ovum space following fertilization.

As mentioned earlier the nonliving universe is a proentropy universe, i.e., its entropy or degree of disorderliness incessantly increases following the birth of regular energy and regular matter.

Generation of dark energy and dark matter, leading to formation of space-time and infinity wall, is followed by generation of regular energy and regular matter as immediate and late derivatives of the sole force or preforce of big bang respectively.

These early and late events are the most fundamental events of nonliving universe.

As mentioned before, the first two are immediate derivatives of big bang and the latter two come into existence following reflection of the residual force of big bang off the infinity or Afrasiabi wall.

Residual force of big bang could best be defined as a force, which is not strong enough for further generation of space-time.

This leads to generation of infinity wall or outermost border of the known universe spontaneously.

However, this residual force is strong enough to give rise to regular energy and regular matter following reflection off the infinity wall.

Thus, infinity wall of the known universe is born because of the relative weakness of residual force.

Just as the regular energy and regular mass are born because of its relative strength.

The condensation of the regular energy in post big bang phase or in other words in big bang residual phase leads to the generation of regular matter in the form of superstrings.

This breakdown of orderliness and symmetry is a never-ending process. Among the major and well-known products of this breakdown are the major forces of electromagnetic, weak nuclear, and strong nuclear forces.

Gravity is the result of indentation of space-time by objects that carry mass rather than a true force.

It could be considered a byproduct of break-down of symmetry rather than a true product.

Finally, as has been well defined irregularities in distribution of matter lead to formation of clumps of matter, which act as the seeds for generation of galaxies.

In living universes with higher level of complexity, similar sequence of events following fertilization ensue as well.

Morula comes into existence and gives birth to the symmetrical blastula.

The symmetry of blastula gets broken down and gives rise to gastrula.

Further breakdown of symmetry leads to formation of three layers of cells, namely ectoderm, mesoderm, and endoderm.

Through further breakdown of symmetry, each layer acquires a different assigned job for generation of different organs of the living universes which are Akin galaxies of the known non-living universe.

Again, the main distinguishing feature between living and nonliving universes is their fundamentally different programming as far as their alignment along the thermodynamics arrow of time.

As said before, the degree of decrease in entropy in living universes reaches a limit, which is set by the second law.

This limit, which acts as a firewall or infinity wall of the living universes, leads to aging, disease, and death of living universes.

By the same token and in contrast, the degree of increase in entropy of known nonliving universe also reaches a limit set by the infinity wall.

This would finally trigger the process of big crunch following maximum expansion of known universe, which leads to demise of known nonliving universe.

The dark part of nonliving universe, which is 96 or so percent of known nonliving universe and is made of dark matter and its quantum dark vibrations or dark energy remains static following its original formation.

Expansion and crunch, which is discussed commonly in scientific literature, apply to regular matter, comprising around 4 percent of the known nonliving universe.

Thus, Albert Einstein was right when he said the universe is static and does not expand, if he really meant dark energy and dark matter part of nonliving universe, which indeed act as the container or p-brane in which regular universe expands and finally goes into big crunch.

In closing, a handful of universal laws thrown in by great grand designer create the foundation of behavior of universe at all levels.

This could best define the logic of living and nonliving universes. This logic prevails throughout the lifetime of the universe.

As such, opposing forces are best defined as illogical. They trigger catastrophic events and a process that leads to their ultimate demise.

CHAPTER 5 ADDITIONAL READING

Davies, P. C., E. Rieper, and J. A. Tuszynski. 2013. Self-organization and entropy reduction in a living cell. Biosystems 111(1), 1–10.

Recordati, G., and T. G. Bellini. 2003. A definition of internal constancy and homeostasis in the context of non-equilibrium thermodynamics. Experimental Physiology 89:1, 27–38.

Sherr C. J. 1996. Cancer Cell Cycles. Science 274:1672–1677.

Wald, R. M. 2001. The Thermodynamics of Black Holes. Living Rev Relative, 4(1), 6.

CHAPTER 6

Puzzling Sensors of Living and Nonliving Universes

The sensors of living and nonliving universes act as guardians of fundamental and prevailing universal laws.

As such, perception of microenvironmental, environmental, and universal cues with proentropy and antientropy properties plays a central role in living and nonliving universes.

Living universe sensors, communicate with downstream intracellular network satisfying maintenance of cellular network energetics homeostasis.

To achieve this goal, the downstream network of sensors of living universes connects itself to all intracellular compartments.

Such compartments incorporate genome, epigenome, and micro- RNA network as well as gene regulatory machineries such as spliceosomes.

Consequently, proentropy cues capable of increasing intracellular network entropy also affect energy sensors such as G-PCRs by modifying GTP-GDP ratio.

G-PCRs, through their downstream networks such as C-AMP and PI3-kinase, could trigger mitosis.

This is dictated by ratio of GTP over GDP-bound G-PCR. As such, abundance of GDP bound G-PCR is associated with high cellular network entropy.

Following an increase in cellular network entropy, major energy sensors such as G-PCRs and AMP kinase trigger generation of more ATP molecules.

This happens by increasing the pace of oxidative phosphorylation in eukaryotes. Activation of mitosis ensues via disinhibition of cyclin dependent kinases.

A major goal of mitosis is restoration of maximum allowable cellular network free energy.

However, mitosis itself is an energy consuming biological process. Consequently, part of the generated ATP molecules is consumed during mitosis.

In contrast, generation of ATP happens by aerobic glycolysis of nonoxidative phosphorylation subtype in aerobic prokaryotes and anaerobic glycolysis in anaerobic prokaryotes.

Restoration of cellular network free energy is multifaceted. Conversion of low free energy components and molecules of cellular network to their high free energy counterparts is one potential mechanism.

This is an energy-consuming process. Another potential mechanism is replacement of old cells with high network entropy with new generation of cells with significantly lower network entropy.

Restoration of inherent normal energetics status of molecules is the key determining factor of their normal function.

As one example, chlorophyll is a well-known high energy macromolecule. Consequently, it transmits photons rather than absorbing them.

This is the foundation of photosynthesis. However, if quaternary structure of chlorophyll would change to a low energy one due to any ill event, photosynthesis would cease.

Thus, available ATP molecules could be used either to restore its original quaternary high energy structure.

As mentioned above, another alternative is to replace such cells with a new generation of cells with well-balanced chlorophyll energetics.

Hence, the total quantity of available intracellular ATP at any time correlates with cellular network free energy in a linear fashion.

On the other hand, antientropy cues could potentially increase intracellular network free energy beyond the allowable limit set forth by the second law.

This would slow down the pace of ATP formation through downstream cellular energy sensors accordingly and thus would restore the energetics balance.

In other words, the two extremes are equally unacceptable and trigger a shift to midline by corrective measures under normal condition.

In case of nonliving universe, events happen effortlessly because there is inherent alignment with the second law of thermodynamics.

The known nonliving universe expands spontaneously along the thermodynamics arrow of time, with ensuing incessant increase in disorderliness following its birth.

This process seems to have a built-in quantum sensor in the sense that any event favoring increased orderliness faces insurmountable barriers to which it eventually succumbs.

One example is evaporation of black holes through their entropy ring or event horizon.

Another example is explosion of certain stars through supernovas. These processes abort the fierce move against asymmetry and entropy by massive force of gravity.

Such corrective events in nonliving universe restore disorderliness when the opposing forces exhaust and reach the built-in firewall.

Similar built-in sensors exist in living universes as well. Design or preprogramming of the molecular components of living cells toward maintenance of maximum allowable free energy seems to be the main mechanism.

I have coined this built-in mechanism Afrasiabi law of spontaneity.

As one example, components of molecular machinery of mitosis get activated by a decrease in cellular network free energy.

This would imply that quaternary structure of key molecular elements of mitotic machinery such as cyclin dependent kinases are themselves sensitive to variations in cellular network entropy and respond accordingly.

Consequently, any major change in their quaternary structure dictated by different gradations of cellular network entropy could trigger their activity or lack thereof as well as interactions with surrounding molecules accordingly.

This could lead to activation of cell cycle versus its inactivation. In addition, the dynamics of interactions among different components of cell cycle machinery as well as protein-protein interactions are guided accordingly.

Puzzling sensors are spread diffusely among all intracellular compartments and throughout the lifetime of living universes to secure well-balanced energetics and maximum allowable cellular free energy.

Another example, is a shift of paternal DNA imprinting toward maternal imprinting signature. This process starts during embryogenesis, and persists throughout the lifetime of complex living universes.

This process might offer an explanation to loss of Y chromosome with aging, which reflects many more rounds of mitosis.

Any interference with this process might act as major contributory factor to malignant disorders.

Indeed, dysfunctional homologous recombination and dominance of nonhomologous end joining and homeologous recombination in malignant disorders favor preservation of paternal imprinting signature.

Another astonishing example is destruction of paternal mitochondria, which contains a five-thousand-nucleotide-rich circular DNA immediately following fertilization.

This denotes that there are puzzling sensors within the fertilized ovum, which recognize the difference between the paternal and maternal mitochondrial energetics signature and trigger their destruction.

By the same token, there should be other puzzling sensors recognizing physical crowding of fertilized ovum following fertilization.

Crowding increases disorderliness and consequently entropy, which inversely correlates with free energy.

Thus, such sensors secure the homeostasis of the second law of thermodynamics within the boundaries of living universes.

The mechanism of paternal mitochondrial destruction is not well understood. Clearly apoptosis is not the responsible mechanisms.

Hence, another unrecognized sensor might be in play. The instantaneous destruction of paternal mitochondria following fertilization happens as though they have been hit by lightning.

Midline alignment of paternal chromosomes inside the fertilized nucleus might be one reason why they do not share the paternal mitochondrial DNA destiny.

There is also a possibility that nuclear membrane is shielding the nonmitochondrial paternal DNA.

In summary, physical crowding and the difference in maternal and paternal genetic imprinting signatures and energetics, contribute to increase in cellular network entropy following fertilization.

These two major factors act as the mechanism of initiation of mitosis.

Energetics signature could be considered the most fundamental signature in the fabric of living universes simply because it shapes the matrix of living universes and guides its fundamental processes throughout their lifetime.

Hence, it is not surprising that breakdown of cellular network energetics homeostasis through breakdown of its sensors opens the door on malignant transformation.

Comparison of living and nonliving universes, also brings to light a lot of surprising similarities.

One example is that only 4 percent of human nuclear genome is used for coding proteins or for protein production through rRNA and tRNA.

The 96 percent of noncoding DNA could be considered the dark matter of living universes. Surprisingly enough, dark matter also comprises 96 percent of nonliving universe matter.

In addition, regular matter from which everything that is known to us is made comprises 4 percent of known matter in the universe.

Again, another astonishing similarity between the ratio of regular matter and dark matter in nonliving universe at one end and the ratio of coding and noncoding DNA in complex living universes at the other end.

These seemingly junk DNA sequences contain paternal and maternal signatures and are used in DNA fingerprinting to identify individual's origin.

In addition, they represent the bulk of maternal and paternal energetics signature or cellular network entropy.

More evidence shows the existence of regulatory elements in these DNA sequences, which are involved in cell proliferation, development, and cancer.

The fact that cellular network energetics guide fundamental events, such as proliferation, development, cancer, and even evolution itself should not be surprising simply because the very existence of the living universe is deciphered by bioenergetics.

As mentioned before, it is quite fascinating that following each round of mitosis, the paternal DNA imprinting signature also becomes more maternal.

As alluded to, this happens through mechanisms such as homologous recombination and inhibition of other mechanisms such as non-homologous end joining (NHEJ) and homoeologous recombination.

The last two mechanisms are employed by pathological conditions such as neoplastic disorders and are suppressed under normal condition.

Meanwhile the mitochondrial genetic signature remains unchanged, simply because it carries a pure maternal energetics signature.

Major disorders such as muscular dystrophy and some neoplastic disorders are associated with perturbances of maternal mitochondrial DNA signature which translate into distorted energetics.

This is frequently manifested as decreased ATP production by mitochondria and activation of Warburg's effect.

This translates into decrease in cellular network free energy, which inversely correlates with cellular network entropy.

As mentioned above, activation of pathologic double strand and single strand DNA break repair mechanisms such as nonhomologous end joining as well as homoeologous recombination, which promote paternal energetics signature are also associated with catastrophic disorders such as cancer.

Absence or inactivating mutations of the genes responsible for physiological single and double strand DNA break repair such as PARP, BRCA1, and BRCA2 lead to the same outcome.

The common denominator among the above scenarios and examples is enrichment of the double helix DNA with paternal imprinting signature.

Such enrichment is perceived as chaos or disorderliness in the fabric of the cell.

Chaos or disorderliness is a manifestation of increased cellular network entropy.

This acts as the main predisposing factor to a vast array of neoplastic disorders, the hallmark feature of which is an increase in cellular network entropy.

Among the other pathological conditions in which the nonhomologous end joining and homeologous recombination mechanisms of double strand DNA break repair take over are neoplastic disorders originating from mismatch repair defective machinery.

Distorted cellular network energetics signature manifested as increase in cellular network entropy could be identified in such cases as well.

The hallmark feature of this increase in entropy is its irreversible nature. This is the key driving force of neoplastic disorders.

It is quite bewildering that embryogenesis, which gives rise to the birth of complex living universes results from the convoluted pathway of mitosis following fertilization.

Such might not have necessarily been the main purpose of recurrent bouts of mitosis following fertilization.

Rather, the main goal of recurrent bouts of mitosis following fertilization is minimizing cellular network entropy.

However, embryogenesis and the birth of complex living universes could be considered the most meaningful quantum product of recurrent bouts of mitosis following fertilization.

The asexual replication of unicellular organisms results from direct and constant penetration of proentropy forces in the immediate environment through their cell membranes and at times, through ill internal events.

The goal of such replication is restoration of maximum allowable cellular network free energy. This process eventually drives evolution to higher levels of complexity.

Thus, one might reasonably conclude that the foundation and fabric of life and its evolution as we know it are intertwined and entangled with its anti-entropy features at its heart.

As mentioned before, in contradistinction to living universes, the driving force for the birth and evolution of the known nonliving universe is its alignment with the thermodynamics arrow of time.

This translates into an ever-increasing pace in universal entropy and its expansion.

The nonliving universe expands spontaneously because of its alignment with the thermodynamics arrow of time.

Along this path, generation of diversity and, in other words, breakdown of symmetry by the derivatives of the force of big bang ensues.

Surprisingly enough, generation of diversity and breakdown of symmetry through recurrent bouts of mitosis, during embryogenesis of complex living universes follows the same path.

One of the most important implications of these concepts is deep understanding of pathological conditions that affect these two different universes.

Living universes cannot afford to succumb to these disorders in the future the way they do today.

Such future is characterized by the need to survive and afford long space travels and the necessity of cloning other planets.

Strong and healthy complex living universes who could continue to work under harsh conditions and for much longer life spans are needed for the next phase of evolution.

As such, disorders such as cancer should become curable and turn into something of the past, just as disorders such as poliomyelitis are today.

Another important implication of these concepts is clear understanding of the faith of living and nonliving universes.

Accordingly, the known nonliving universe could not continue to expand forever as is currently perceived by scientific society.

Rather it would go into big crunch after its expansion reaches the infinity or Afrasiabi firewall of the known nonliving universe.

Based on these concepts and parallelism of events of living and nonliving universes, one could envision that in nonliving known universe, situations like neoplastic disorders of the living universe should also exist.

Accordingly, there should be regions of space-time in which the thermodynamics arrow of time is frozen or is going the opposite way.

In such regions, time stops from running and so does increase in entropy, and one could potentially travel to the past.

Such regions of space-time could represent malignant transformation of that section of non-living universe, simply because they oppose the ever-increasing pace of entropy.

This is akin the behavior of a malignant cell, which opposes the significant decrease in entropy of the living universe.

Black holes seem to represent such aberrancies. In such regions of the known nonliving universe time comes to an end and thus entropy shifts toward zero.

However, black hole entropy cannot reach zero, just the way entropy cannot reach zero in living universes because of the limits set by the second law.

In this case, virtual particles with negative mass coming into existence in the event horizon of black hole fall into black hole and gradually decrease its mass.

This evaporation of black hole mass is visualized as emission of virtual particles with positive mass which can run away from black hole.

As such, black hole acquires entropy represented by radiation from its event horizon.

The bigger the surface area of the event horizon of the black hole, the larger the amount of radiation emitted from the black hole. This, translates into higher black hole entropy value.

Thus, devouring more and more of the surrounding matter by black holes and increasing the circumferential diameter of their event horizon, acts like a double-edged sword in the quantum universe.

At one end, it leads to the growth in size of the black hole and at the other end it promotes its demise as mentioned above.

As it has been well described in physics literature, black hole demise happens through loss of mass originating from its event horizon radiation.

Throughout this process, black hole becomes smaller and smaller and hotter and hotter.

It finally explodes and its demise leads to generation of new galaxies or even other nonliving universes, which obey the alignment along the thermodynamics arrow of time.

This is like cancer, which devours substrate or matter and energy resources of its living universe host.

This process also leads to further growth and metastatic spread of cancer, which eventually leads to demise of the living universe host and cancer.

Recycling of the matter comprising the demised living universe host and tumor mass leads to generation of other living universes, which obey the principles of the second law in living universes.

Again, we see parallelism of events between living and nonliving universes.

In summary, puzzling sensors of living and non-living universes safeguard a handful of laws, which have led to their birth and evolution.

As such, no aberrancy would go unrecognized and demise of the aberrant entity disobeying the prevailing laws, be it a black hole in nonliving universe or cancer in living universes, secures the integrity of the laws prevailing these universes.

CHAPTER 6 ADDITIONAL READING

Afrasiabi, K. 2011. Entropyomics as the Blueprint of the Logic of Normal Cell Division and Malignancy. Online Journal of Biological Sciences 11(1), 23–26.

Afrasiabi, K., Y. Zhou, and A. Fleischman. 2016. GPCRs, at the Crossroad of Distortions in Extracellular Microenvironment and Intracellular Energetics Homeostasis, a New Model for 21st Century Cancer Therapeutics. Cancer and Clinical Oncology, 5:1.

Ambekar, S. S., S. S. Hattur, and P. B. Bule. 2017. DNA: Damage and Repair Mechanisms in Humans. Glob J Pharmaceu Sci 3(2).

Czihak, G. 2012. *The Sea Urchin Embryo: Biochemistry and Morphogenesis.* Springer Science and Business Media.

Hardie, D. G. 2011. Sensing of energy and nutrients by AMP-activated protein kinase. Am J Clin Nutr, 93:4. 891S–896S

Keveme, E. B. 2015. Genomic imprinting, action, and interaction of maternal and fetal genomes. Proc Natl Acad Sci USA 112(22), 6,834–6,840.

Lawson, H. A., J. M. Cheverud, and J. B. Wolf. 2014. Genomic imprinting and parent-of-origin effects on complex traits. Nat Rev Genet *14(9):* 609–617.

Lieber, M. R., J. Gu, H. Lu, N. Shimazaki, and A. G. Tsai. 2010. Nonhomologous DNA End

Joining (NHEJ) and Chromosomal Translocations in Humans. Subcell Biochem. 50, 279–296.

Schon, E. A., S. DiMauro, and M. Hirano. 2012. Human mitochondrial DNA: roles of inherited and somatic mutations. Nat Rev Genet, 13(12), 878–890.

Traschen, J. 2000. An Introduction to Black Hole Evaporation. World Scientific 2000, 180.

Zhou, Q., H. Li, H. Li., A. Nakagawa, J. Lin, E. Lee, B. Harry, R. R. Skeen-gaar, Y. Suehiro, D. William, S. Mitani, H. S. Yuan, B. Kang, and D. Xue. 2016. Mitochondrial endonuclease G mediates breakdown of paternal mitochondria upon fertilization. Science *353(6297)*. 394–399.

CHAPTER 7

The Conscious Living and Nonliving Universes

Consciousness might be best defined as a state of awareness encompassing the processes of initiation and pursuit of a path in a goal oriented fashion.

Living universes constantly generate and spend energy to maintain minimum amount of cellular network entropy.

Thus, the goal of living universes could be defined as maintenance of minimum allowable cellular network entropy, which correlates with maximum achievable cellular network free energy.

In contrast, in nonliving universe there is incessant increase in disorderliness following its birth.

In addition, there is entanglement between its spontaneous expansion along the thermodynamics arrow of time and incessant increase in universal disorderliness.

Such persistent increase in disorderliness could be considered its goal.

In contrast to living universes where there is constant need for generation and consumption of energy to achieve their goal to maintain minimum amount of cellular network entropy, in nonliving universe, such goal is achieved effortlessly because of its alignment with second law of thermodynamics.

In living universes, to ensure that such goal is achieved, a state of awareness has been programmed into their fabric.

Such programming encompasses intracellular communication network and extends to all subcellular compartments ranging from genome to RNA and protein expressions and protein-protein interactions.

As such, in living universes, any deviation from minimum allowable cellular network entropy set by the limits of the second law triggers multiple sensors to secure this principle.

As one example, any change in cellular network energetics spontaneously changes the quaternary structure of the components of intracellular communication network.

This change would trigger a series of events including a change in protein-protein interactions.

Any deviation in cellular network landscape energetics also affects other intracellular compartments such as genome, epigenome, microRNA network, and spliceosomes.

The resulting change in structure, function and dynamics of the complex interactions among such compartments and components, decide about activation or deactivation of the mitotic and ATP generating machineries.

Hence, volcanic eruptions of mitosis could be considered an attempt at restoration of minimum allowable cellular network entropy.

Simply because the born daughter cells with inherently lowest possible network entropy would survive, and the progenitor cells with high network entropy would succumb through different mechanisms such as apoptosis or autophagy.

The same state of goal-oriented awareness is also programmed into the fabric of the microenvironment.

In this regard, any perturbations in the energetics of the microenvironment would get conveyed to neighboring cells.

This would trigger the same cellular sensors and pathways toward restoration of balance and harmony between cellular and microenvironment energetics.

This demands a delicate and meticulously programmed interconnectivity between microenvironment and adjacent cells.

Extracellular matrix proteins seem to have acquired the evolutionary task of communication between these two critical compartments of living universes.

In this regard, fibulins could be considered important members of such matrix proteins.

This is best demonstrated by the fact that fibulins as well as other EMP members are dysfunctional in nearly all cancers.

It seems quite likely that state of awareness of energetics of microenvironment is conveyed through physical cues of its matrix proteins.

To achieve this goal, the status of microenvironment energetics should affect a multitude of physical features including the quaternary structure of its matrix proteins.

Additionally, it would be desirable that such matrix proteins could reciprocate and affect the energetics of the microenvironment.

Thus, it is not surprising that the unique feature of fibulins is intramolecular bridge formation of proteins comprising the matrix.

Accordingly, it is plausible that folding of matrix proteins by fibulins modifies the physical features of microenvironment, which correlate with its energetics status.

For example, a folded matrix protein lowers free energy of microenvironment as compared with its relaxed conformation.

This happens simply because energy is consumed for formation of intramolecular bonds and the process of folding.

The final piece of this delicate and orchestrated programming of awareness should do with conveyance of variations of the energetics status of the microenvironment to the cellular network and vice versa.

This seems to happen in part through variable cellular expression and release of EMPs such as fibulins (EFEMP1, FBLN5) and their regulated matrix metalloproteinases.

Breakdown of such balance leads to the birth of malignant phenotype, leads to activation of oncogenic signaling pathways in cancer cells, and cancer-specific metabolism, including aerobic glycolysis.

Aerobic glycolysis pathway in cancer cell is also known as Warburg's effect.

This would eventually lead to an irreversibly low cellular network free energy, which is the hallmark of malignancy.

Eventually, a reverberating vicious cell cycle ensues, which adds more depth to the decrease in cellular free energy and landscape energetics.

Consequently, irreversible ill events ensue following breakdown or deprogramming of sensors supposed to secure the state of awareness built into the fabric of these essential and intimate compartments.

This could further consolidate the significant decrease in cellular network free energy through overexpression of NOTCH relative to EGFR pathway in stem cell compartment.

Overexpression of NOTCH is among the essential factors that establish and perpetuate stemness identity.

By now, glycolysis has become the dominant metabolic path and oxidative phosphorylation has lost its dominance.

Aerobic glycolysis leads to a significant decrease in net ATP production capability, which perpetuates the decrease in cellular network free energy.

Persistent decrease in cellular network free energy could irreversibly affect the energetics of critical compartments of the cell.

Such compartments include, but are not limited to, epigenome, genome, RNA spliceosomes, microRNA network, and protein-protein interactions.

GPCRs, which play a critical role as energy sensors, have also been shown to get mutated or dysfunctional in a significant percentage of malignant disorders.

Because of perpetuation of such major and irreversible perturbances, full-blown malignant phenotype would ensue.

Breakdown of a very intimate relation and cross talk between cell and microenvironment is another consolidating factor of malignant transformation.

There are probably other potential mechanisms that work toward maintenance of energetics homeostasis that could also break down during the process of malignant transformation.

As one example under normal condition, low free-energy microenvironment could potentially filter out molecules or factors that could lead to further decrease in free energy of microenvironment and consequently the cell.

This could represent another major programmed sensor mechanism of microenvironment toward maintenance of the energetics homeostasis of both compartments that could fall apart.

This breakdown could potentially lead to a different set of pathologic interactions with the outside environment.

The resulting interference with restoration of energetics homeostasis of microenvironment and consequently the cell, further complicates the situation.

This could culminate in energetics crisis of the cell, and eventually irreversible deprogramming of cellular energetics homeostasis sensors and a bigger shift toward malignant transformation.

Our future cancer treatment modalities should underpin the location, nature, and depth of these pathologies and generate customized treatment strategies accordingly.

Thus, in summary the existence of fine-tuned and goal-oriented interactions between the microenvironment and the cell as described above further defines and solidifies the concept that the living universes are indeed conscious.

On a different note, nonliving known universe, in contradistinction to living universes, is aligned with the thermodynamics arrow of time.

Nonliving universe events that happen spontaneously following its inception and are left uninterrupted and undisturbed are those that contribute to increase in the degree of disorderliness.

Throughout this process, free energy of the closed and infinitely vast nonliving universe continues to decrease spontaneously.

In contrast, in living universes energy needs to actively and continuously get generated and consumed efficiently to maintain orderliness and mini-mize network entropy.

Breakdown of homeostasis in living universes happens when such mechanisms fall short of achieving such goal.

In nonliving universe, such breakdown happens when an event translates into an increase in the degree of orderliness.

The living universe is conscious because it constantly monitors and secures maintenance of the highest possible network free energy through its built-in sensors.

In contradistinction, the nonliving universe is conscious because it guides and secures an incessant and uninterrupted rise in universal disorderliness or entropy.

Any ill event that escapes the sensors of these two universes, which act as guardians of the prevailing rules, eventually leads to demise of that ill event.

If the ill event is big enough to extend to other compartments of that universe, the extent and scope of demise would increase accordingly.

Cancer in living universes serves as one such example. Consequently, it ultimately leads to premature demise of its living universe host.

Black holes and certain rotating galaxies that oppose alignment along the thermodynamics arrow of time are good examples of ill events in non-living universe.

In black holes, time comes to an end because massive gravity of black hole ceases all kinds of motion including that of photons, the motion of which is the generator of time.

In rotating galaxies and probably some baby universes, motion is along an axis perpendicular to thermodynamics arrow of time.

Consequently, imaginary time, which is perpendicular to real time axis, becomes dominant and real time comes to an end.

The common denominator that leads to demise in nonliving universe is defiance of thermodynamics arrow of time. Such trapped and doomed clusters of matter could be planets, stars, or galaxies.

The same destiny is awaiting the whole known non-living universe, as it hits the infinity or Afrasiabi wall, at which point thermodynamics arrow of time freezes.

This freeze of thermodynamics arrow of time could also be considered as defiance. Time comes to an end and the processes of big crunch starts.

The natural history of a black hole since the time of its inception to its demise, offers solid evidence to the conscious nature of the known non-living universe.

In summary, by the nature of its definition, black hole incarcerates photon and freezes time.

The passage of time in the known nonliving universe and the thermodynamics arrow of time are aligned along the same axis.

As this alignment ultimately aligns with increase in universal entropy, a black hole is evading this fundamental principle.

Conscious nonliving universe is cognizant of this evasion. The demise of black holes is an attestation to this notion.

This could be a very lengthy process. On occasion, it could take trillions of years, but it ultimately happens.

The process of elimination of black hole starts in its event horizon, which is the outer most rim of a black hole.

Indeed, growth of black hole is also represented by increase in surface area of the same zone.

That would inferentially mean that that the seed of growth of black hole is its suicide machinery as well.

The constant birth of virtual particles with positive and negative energy which is said to happen in all regions of the universe, also happens near the event horizon of the black hole.

Such virtual particles usually collide and get annihilated. Evidently near a black hole the particle with negative energy falls into the black hole.

The particle with negative energy carries a negative mass. This gradually decreases the mass of the black hole.

The particle with positive energy is said to escape the black hole. This escape is a manifestation of radiation from an evaporating black hole.

Thus, the conscious nonliving universe protects and secures its most fundamental law, namely the second law of thermodynamics.

This happens by recognition and elimination of evading antientropy events at one hand and promotion as well as perpetuation of concordant and aligning events.

In contradistinction, in living universes the programmed state of awareness is in recognition of proentropy events.

Such events have shown weakness as far as the power of minimizing the speed of rise in entropy is concerned.

The conscious living universes handle these events in different ways, ranging from repair and corrective measures to apoptosis of their cellular units.

If such measures fail, the conscious living universes endorse their own demise to eliminate the ill events simultaneously.

As such, a patient with metastatic cancer dies and so does metastatic cancer, which has evaded the fundamental rules of birth and evolution of living universes.

Similarly, the conscious old living universes also die because the inevitable piled up entropy down the thermodynamics arrow of time would make life sustaining processes impossible.

Their death leads to recycling of their building blocks and the birth of new healthy conscious universes.

CHAPTER 7 ADDITIONAL READING

Bar-Shavit, R., M. Maoz, A. Kancharla, J. K. Nag, D. Agranovich, S. Grisaru-Granovsky, and B. Uziely. 2016. G Protein-Coupled Receptors in Cancer. International Journal of Molecular Sciences, *17(8),* 1,320.

Chen, P., and G. Mourou. 2017. Accelerating Plasma Mirrors to Investigate the Black Hole Information Loss Paradox. Phys. Rev. Lett. *118(4).*

Gallagher, W. M., C. A. Currid, and L. C. Whelan. 2005. Fibulins and cancer: friend or foe? Trends Mol Med *11(7),* 336–40.

Kayser, K., S. Borkenfeld, R. Carvalho, and G. Kayser. 2015. The concept of entropy in histopathological diagnosis and targeted therapy. Diagnostic Pathology.

Kompanichenko, V. 2017. *Thermodynamic Inversion: Origin of Living Systems.* Springer International Publishing.

Marín, D. and B. Sabater. 2017. The cancer Warburg effect may be a testable example of the minimum entropy production rate principle. Phys. Biol. *14(2).*

Metze K., R. L. Adam, G. Kayser, K. Kayser. 2010. Pathophysiology of Cancer and the Entropy Concept. In: Magnani L., Carnielli W., Pizzi C. (eds) *Model-Based Reasoning in Science and Technology. Studies in Computational Intelligence,* vol *314.* Springer, Berlin, Heidelberg.

Müller, D. J., N. Wu, and K. Palczawski. 2008. Vertebrate Membrane Proteins: Structure, Function, and Insights from Biophysical Approaches. Pharmacological Reviews, *60(1)*, 43–78.

Trevors, J. T., and M. H. Saier. 2011. Thermodynamic perspectives on genetic instructions, the laws of biology, diseased states and human population control. Comptes Rendus Biologies, *334(1)*, 1–5.

Teschendorff, A. E., and S. Severini. 2010. Increased entropy of signal transduction in the cancer metastasis phenotype. BMC Systems Biology, *4*, 104.

Wang, E., A. Lenferink, and O'Connor-McCourt. 2007. Cancer systems biology: exploring cancer-associated genes on cellular networks. M. Cell. Mol. Life Sci. *64*: 1,752

CHAPTER 8

The Blueprint

All events happen on the face of known nonliving universe, or p-brane as addressed in contemporary physics.

The rules that govern and decide about those events cannot obviate the prevailing rules of known nonliving universe which have also decided about its very existence.

In this chapter, I would examine various seemingly unrelated events, and I would conclude that they follow the same universal blueprint.

The implications of such analysis and understanding would be diverse and extend through all fields of science.

This blueprint would enable us to come up with the true meaning of these events.

Such understanding would also empower us with the capability of designing new methodologies to solve puzzles that have surrounded us for ages.

In this regard, the enigma surrounding the mechanism of initiation of mitosis in complex living universes and cell division in simple living universes could serve as a starting and important example.

Our biggest handicap at decoding the puzzle of cancer is lack of deep understanding of logic of normal cell division.

Our generation has spent a lot of time and energy at deciphering the puzzle of cancer and its treatment.

This has taken us from chemo and radiation therapy to targeted and immunotherapy in the last seventy-five years.

Cure is not possible unless we come up with a clear understanding of normal state and the mechanism of initiation of normal cell division in living universes.

As mentioned in earlier chapters, living universes are like fish in the vast ocean of known non-living universe and understanding of fish necessitates deep understanding of ocean.

Over the last several decades, we have gathered vast amount of detailed information about intracellular communication network, as well as its different components, pathways, and molecules involved in cell division.

The spectrum of our knowledge extends from cyclin-dependent kinases and their positive and negative control feedback loops, to mitotic spindle apparatus, Aurora kinases, trans-membrane kinases, and telomere biology.

TP53 and BCL2 families, as well as microRNA network, and epigenome are among other areas that our knowledge has enhanced significantly in the last several decades.

Our growing knowledge about epigenome and microRNA network, as well as our understanding of gene regulatory machinery and RNA spliceo-some complexities has shed more light on the way we look at cellular events.

In a nutshell, all these compartments act in an orchestrated and coordinated fashion and are connected to cell division machinery and contribute to homeostasis of living universes.

Mitosis, while acting as the main engine to generate off spring cells, also is an ingenious machinery that maintains maximum allowable cellular free energy which correlates with the lowest entropy.

Thus, as mentioned before in living universes, mitosis is the main antientropy machinery.

As such, the distinguishing feature between living and nonliving universes is their opposite alignments along the thermodynamics arrow of time.

In this sense, the nonliving universes are entropy generating machineries while living universes are antientropy engines.

Another astonishing example in complex living universes is meiosis in their gonads.

Throughout this process, haploid sperms and ova come into existence. This is transiently preceded by generation of unstable hyper-diploid cells.

However, haploid cells cannot propagate without fertilization, even though their network entropy is significantly less than diploid cells, which is the most desirable feature in living universes.

This is an attestation to the fact that the goal of mitosis is decrease in cellular network entropy and in haploid cells saturation point has been reached.

This also denotes, that in quantum living universes there is entanglement of low and high network entropy systems. In other words, the path of one passes through the other.

That is to say, haploid cells or germ cells arise from complex diploid living universes and vice-a-versa.

On a different note, the astonishing parallelism between living and nonliving universe events could be seen at each point in their evolution.

As one example, the volcanic eruption of mitosis in complex living universes following fertilization represents a quantum shift toward lower cellular network entropy.

The resulting quantum cellular network landscape energetics would represent a balanced system that lends itself to evolution of living universes to higher level of organization and complexity.

However, unicellular organisms also undergo spontaneous and incessant rounds of cell division. In this case, fertilization is nonexistent.

Incessant penetration of unicellular organism barriers by proentropy forces of the surrounding environment seems to be the trigger for cell division.

Thus, the driving force on both occasions seems to follow the same principle or blueprint, which pushes and keeps cellular network entropy at minimum allowable level as per the limits of the second law.

In contrast to living universes, in known non-living universe the blueprint pushes for ever-increasing entropy and the main engine for creation of space-time is big bang.

The stability of space-time originates from entanglement of dark matter and dark energy, which are the immediate derivatives of big bang.

Dark energy, which is the probable quantum dark vibration of the wave function of dark matter, stabilizes dark matter scaffold of space-time through entanglement with dark matter identity.

This scaffold acts as the birthplace of galaxies, which as mentioned before, have come into existence in post big bang phase.

In this sense, entanglement of dark matter and dark energy resembles fertilization of ovum by sperm.

In other words, following entanglement or fertilization of ovum by sperm a new and stable cell or living universe comes into existence.

This event establishes the foundation of complex living universes and triggers incessant rounds of mitosis.

This results in generation of organs of the new living universes, akin galaxies of known nonliving universe.

The universal event that could best define fertilization is sudden and massive increase in entropy, which results from physical crowding and biological disparities.

In other words, population and biological diversity of chromosomes increase instantaneously following fertilization in a closed space previously occupied by ovum.

The resultant decrease in freedom of motion in the limited space supplied by ovum, translates into a decrease in free energy, which inversely correlates with entropy.

In addition, the paternal genetic imprinting is totally different from the maternal imprinting pattern.

Thus, fertilization would also impose an increase in unwanted biological diversity which translates into further increase in the index of disorderliness.

As such, index of instability, which correlates with entropy, would increase further.

This would act as an additional and separate force, pushing for recurrent rounds of mitosis to minimize entropy.

As mentioned before, paternal genetic imprinting also becomes more maternal following each round of mitosis after fertilization.

Thus, recurrent rounds of mitosis during embryogenesis, respect the blueprint and move toward achievement of this goal.

In other words, instantaneous physical and biological crowding and dissimilarities in the closed ovum space act as an insult to the blueprint of living universes.

This event seems to be the main mechanism of initiation of mitosis, the main goal of which is restoration of integrity of the blueprint.

In a rather similar way, original generation of dark energy and dark matter and consequent generation of regular energy and regular matter are derivatives of the force of big bang and residual force of big bang respectively.

As such, the orderliness and symmetry is disrupted and the force of big bang gets broken down into multiple derivatives each with a different alignment and assigned job.

Irregularities in distribution of matter lead to formation of clumps of matter, which act as the seeds for generation of galaxies.

All these events are respectful of the blueprint of nonliving universe which dictates incessant increase in disorderliness.

Surprisingly enough, following fertilization similar sequence of events ensue. Morula comes into existence and gives birth to blastula.

The symmetry of blastula gets broken down and gastrula comes into existence.

Three layers of cells, namely ectoderm, mesoderm and endoderm evolve following formation of gastrula.

In this case, breakdown of symmetry represents increase in orderliness, respecting the blueprint of living universes.

Each layer has a different assigned job, leading to formation of different organs of living universes which are akin to galaxies of known nonliving universe.

As mentioned before, the main distinguishing feature between the blueprint of living and nonliving universes is their different alignment along the thermodynamics arrow of time.

As said earlier, the degree of decrease in entropy in living universes reaches a limit set by the second law.

This limit which acts like a firewall leads to aging, disease, and death of living universes.

By the same token the degree of increase in entropy of the known nonliving universe also reaches a limit set by the infinity or Afrasiabi wall.

This would trigger the process of big crunch following maximum expansion of known non-living universe which eventually leads to its demise.

It is not clear if the big crunch happens at the edge of the infinity wall, or there would be a repulsive force at that border taking the galaxies backward and toward a central point before crunching them

In contrast, in living universes, energetics signature could be considered the most fundamental presentation of the blueprint.

This is simply because the very existence of living universes is deciphered by bioenergetics principles.

Thus, it should not be surprising that following each round of mitosis, the paternal nonmitochondrial DNA landscape energetics signature becomes more maternal.

One needs to keep in mind that the dominance of maternal DNA energetics signature reflects a decrease in cellular network entropy or disorderliness simply because paternal energetics signature reflects additional unwanted diversity or disparity.

Nonhomologous end joining as well as homeologous recombination following single and double strand DNA break are dominant repair mechanisms in certain neoplastic disorders.

This would enrich the double helix with paternal DNA sequences, which in turn would lead to a significant increase in cellular network entropy, giving further depth to neoplastic transformation.

These two mechanisms of double and single strand DNA break repair take over in conditions with defective homologous recombination.

The convoluted pathway of mitosis following fertilization is aiming at decrease in entropy.

Such decrease in disorderliness is seen in orderly compartmentalization and organization of the embryo at one end, and maternal cellular energetics signature dominance at the other end.

Embryogenesis, and birth of a new living universe from volcanic eruption of mitosis following fertilization, is simply astonishing.

Thus, one might reasonably conclude that the foundation and fabric of life are intertwined and entangled with its antientropy feature or blueprint.

In contrast, the driving force or blueprint for expansion and evolution of the known nonliving universe is its alignment with the thermodynamics arrow of time.

Thus, the known nonliving universe expands spontaneously because entropy increases with expansion.

In this regard, expansion in nonliving universe and mitosis in living universes could be considered as the engines turned on by the blueprints of these universes.

One of the most important consequences of breakdown of living universe blueprint is the rise of aberrancies such as neoplastic disorders.

These disorders are characterized by incessant rounds of dysregulated mitosis and a significant and inappropriate increase in the malignant cell entropy as demonstrated in an elegant fashion by published mathematical models.

Thus, uncontrollable and dysregulated rounds of mitosis in neoplastic disorders seem to be a futile attempt at decreasing entropy.

In known nonliving universe, conditions like neoplastic disorders of living universe exist.

Such conditions or states are best exemplified by regions of space-time in which the thermodynamic arrow of time is frozen.

Black holes are among such aberrant examples. The natural history of many black holes is devouring more and more of the surrounding matter and increase in the circumferential diameter of their event horizon.

This resembles cancer, which devours matter and energy resources of the living universe host.

On both occasions, blueprint is insulted and the aberrant events lead to their own demise.

Thus, it seems quite plausible that virtual particles with negative mass and energy near black hole fall into the black hole and the virtual particles with positive mass and energy run away to infinity.

Consequently, the black hole continues to lose mass in the form of Hawking radiation and shrinks in size and increases in temperature.

Finally, massive explosion of extremely hot and dense and much smaller in size black hole happens.

The life cycle of black hole comes to an end and its residual gets recycled into new universes or galaxies, which obey the blueprint by aligning themselves with the thermodynamics arrow of time.

By the same token, the life cycle of neoplastic disorders also comes to an end following metastatic spread or progression of disease which leads to demise of host.

The decayed matter of the demised host gets recycled into other new living universes, which obey the blueprint or alignment against the thermodynamics arrow of time.

As such, the blueprint cannot be obviated and continues to be respected by both the living and nonliving universes.

CHAPTER 8 ADDITIONAL READING

Odde, D. J. 2015. Mitosis, Diffusable Crosslinkers, and the Ideal Gas Law. Cell 160(6), 1,041–1,043.

Pento, J. T. 2017. Monoclonal Antibodies for the Treatment of Cancer. Anticancer Research: International Journal of Cancer Research and Treatment. 37(11), 5,935–5,939.

Seyhan, A. A. 2016. A multiplexed miRNA and transgene expression platform for simultaneous repression and expression of protein coding sequences. Mol Biosyst 12, 295–312.

Van Wieringen, W. N., and A. W. van der Vaart. 2011. Statistical analysis of the cancer cell's molecular entropy using high-throughput data. Bioinformatics 27(4), 556–563.

CHAPTER 9

Quantum Stem Cell in Quantum Universe

We are the apparent residents of quantum universe. As such, all events have quantum identity and quantum features.

Even though our perception and measurements seem certain to us, the truth is that they are based on uncertain grounds.

What we perceive and define as particle is indeed a wave. What we measure as one path traversed by one particle moving from point A to B, is indeed innumerable traversed paths.

In this regard, our perception reflects the most probable path relative to our position in space-time.

What we perceive as the nonliving universe is indeed one of the infinite number of universes.

Our illusionary perceptions and measurements extend to all realms of scientific fields and all aspects of our lives.

As such, we would like to come up with one definition of an event and one conclusion of its outcome.

In quantum universe, there is nothing like one definition and there is nothing like one conclusion.

There are many definitions of one event and many conclusions of one outcome.

Surprisingly enough, they are all equally right, and they are all equally wrong.

Consequently, our current perception and definition of stem cell in biology is not an exception.

Such perception and definition of stem cell has contributed to design of current treatment strategies for disorders such as cancer.

Such treatment strategies have not led to generation of fundamental solution of these disorders yet.

It is only when we develop a quantum perception and definition of stem cell and biological events that we could come up with a revolution in the design of therapeutics in general and cancer treatment specifically.

Hence, it is not convincing that a tiny fraction of cells in each organ should have taken over the homeostasis and maintenance role of that organ in an indispensable and irreversible fashion.

Indeed, high throughput technology-based experiments have proven that an ordinary cell could develop stem cell identity.

This has been witnessed following destruction of malignant stem cells of intestinal crypt and replenishment of the whole crypt population by surviving neoplastic transit amplifying cells.

So far, this is the best attestation to the fact that the identity and features of a cell known to us as stem cell could be acquired by any cell in the population as demanded and dictated by the new condition, as shown in generations induced pluripotent stem cells (also known as iPS cells or iPSCs).

Thus, there is no restriction and exclusiveness for such identity. Indeed, such identity is deciphered by the prevailing quantum features determining the interactions among the comprising cells in the neo-plastic population at one end and microenviron-ment and those cells at the other end.

Examination of some of the features of these cells might shed light on their convertible identity and dispensable nature.

Unlike other cells, these cells are quiescent most of the time; however, they occasionally undergo vol-canic eruptions of mitosis.

As mentioned in previous chapters, this volca-nic eruption of mitosis is the response to sudden increase in cellular network entropy.

Rapid bursts of mitosis are considered an attempt at restoration of minimum allowable cel-lular network entropy, which correlates with maxi-mum allowable cellular network free energy.

In human bowel, this happens every three or so days in the few cells that are at the base of intestinal crypts.

The reason for this timely and regular schedule of volcanic eruption of mitosis remains elusive.

A preprogrammed, self-inflicting chaos leading to sudden increase in cellular network entropy is an appealing hypothesis in this regard.

This could happen through a complex and fine-tuned interaction between those few cells and the adjacent microenvironment.

Circulating noxious agents, capable of increasing the cellular network entropy that could penetrate the barriers of these cells, is another potential explanation.

Finally, the quantum nature of the genetic machinery of those cells could endow upon them the properties of a quantum biological clock, which would periodically increase the cellular network entropy to a level that could activate another round of mitosis.

Based on the above premises, one might reasonably conclude that any cell, which is mitotically silent, has reached the maximum allowable free energy or the lowest possible entropy and any cell that is dividing is trying to achieve that goal.

Thus, the graded spread of universal quantum waves of entropy into the cell population deciphers the identity and role specification of the cells comprising that population.

One of the implications of this concept is the development of future stem cell–based therapeutics.

This would have a diverse range of applications extending from replacement of hematopoietic machinery in disorders such as aplastic anemia to treatment of myelodysplastic syndrome and leukemia.

Treatment of nonmalignant disorders such as myocardial infarction, cardiomyopathies, and auto-immune disorders could also get revolutionized accordingly.

Our future generation of genetic engineers should master quantum programming of any cell and its conversion into stem cell.

Such cell could be isolated from the organ of interest. I would define and propose transfection of quantum super vectors into these cells as a new methodology to achieve this goal.

Quantum super vector is not bound by a restricted and limited sequence of DNA of which it is comprised.

In contradistinction to the currently used vectors that deliver a predetermined and known genetic material into the cell, quantum super vector would deliver multiple waves of genetic information into the same cell.

As such, incorporation of the first wave or sequence by the vector could activate the main genes associated with stem cell identity.

The second wave of the genetic information of quantum vector would get activated by expression of the first wave genes.

Different mechanisms could get employed in this regard. One such mechanism is activation of the second wave by first wave products, such as RNA and protein.

Activation of the second wave of the quantum vector genes would lead to proliferation of the cells with active first wave genes.

After reaching a critical and preprogrammed number of mitosis, the third wave of the quantum super vector genes would become activated which would take these cells into G0 or quiescence.

Thus, we would be able to generate and proliferate cells with stem cell identity.

These ex-vivo produced stem cells could get delivered into the organ of interest through nanodelivery technology.

Upon exposure to the microenvironment of the organ of interest, the microenvironmental molecules such as PDGF, VEGF, or extracellular matrix proteins would be programmed to activate the fourth wave of the quantum super vector genes.

Activation of the fourth wave would lead to volcanic eruption of mitosis and differentiation into daughter cells of interest and repopulate the organ of interest.

In case of aplastic anemia or myelofibrosis, the full hematopoietic repertoire would come into existence again.

In case of neoplastic disorders, one could program multiple waves into the quantum super vectors and then carry them by nanodelivery technology into malignant cells.

Upon incorporation into the malignant cells of the tumor mass of interest, numerous mechanisms of modification or silencing of their malignant genes could get activated.

One way to achieve this goal is through a nanoswitchboard built into the quantum super vector.

Exposure to certain transcription factors or cytoplasmic proteins specific to that type of cancer cell could act as potential mechanisms of activation of the quantum super vector switch board.

Such quantum super vectors could be designed to move out of one malignant cell into another.

Different malignant cells, comprising the intra-tumor heterogeneity, would activate a different nanoswitchboard program of the quantum super vector based on their specific genetic, microRNA, or protein-protein interaction signature.

Activation of different programs in different cells would lead to conversion of that malignant phenotype into their normal counterparts.

A diverse group of over-the-counter quantum super vectors could get designed and produced to modify different sub compartments of the cells of interest in a goal oriented fashion.

Such programmed modifications could extend from cancer cell epigenome all the way to microRNA network in multiple well-coordinated steps and waves.

The future of biomedical sciences would be built by scientists well familiar with this concept and well equipped with the design and delivery technology.

One could envision the future of cancer thera-peutics by dominance of over the counter quantum super vectors.

Each one with a specific bar code designed for treatment of a specific disease, ranging from neoplastic, auto-immune, inflammatory, and congenital disorders.

They would be available to the new generation of biomedical scientists and practicing doctors.

Thus, the underappreciated concept of infinity loop has strong bearing on our current lack of success in eliminating cancer through stem cell targeted therapeutics.

This concept denotes that the quantum stem cell resides in the infinity loop of the quantum universe.

Residence in this loop is dictated by the quantum nature of the stem cell.

A simple way to grasp this concept is through the famous dictum of quantum mechanics.

It is well known and proven in quantum mechanics that particles occupy innumerable or infinite number of positions at the same time.

Stem cells are made up of nothing but such particles or superstring vibrations.

Thus, our perception of their location at a specific position in an organ or organoid is also an illusion.

Consequently, targeting that physical radius with any of the available or designed therapeutics would lead to failure.

Current failing results of such therapeutics are the best attestation to this fact.

In closing, better and deeper understanding of the nature of our quantum universe and consequently quantum cell would enable us to open the door on the bewildering world of quantum biology.

This journey would lead to the birth and development of quantum therapeutics.

As such, revolution of therapeutics in general and cancer therapeutics in specific would be pioneered by the new generation of quantum biologists.

They would design quantum therapeutics for malignant and nonmalignant disorders and would deliver them by technologies such as nanodelivery into the infinity loop.

By walking down this path, we would become able to go for a home run and conquer catastrophic diseases such as cancer and open a new chapter in evolution of living universes.

CHAPTER 9 ADDITIONAL READING

Davies, P., L. A. Demetrius, and J. A. Tuszynski. 2012. Implications of quantum metabolism and natural selection for the origin of cancer cells and tumor progression. AIP Advances, *2*(1),

Feynman, R. P. 1942. The Principle of Least Action in Quantum Mechanics, Ph.D. thesis, Princeton, May 1942.

Robens, C., W. Alt, D. Meschede, C. Emary, and A. Alberti. 2015. Ideal negative measurements in quantum walks disprove theories based on classical trajectories. Phys. Rev. X 5(1).

Rolfe, D.F., and G. C. Brown. 1997. Cellular energy utilization and molecular origin of standard metabolic rate in mammals. Physiol. Rev. *77*, 731–758.

Torday, J. S., and V. K. Rehan. 2012. *Evolutionary Biology: Cell-Cell Communication, and Complex Disease.* New York: John Wiley & Sons.

Mitochondria, the Powerhouse of Living Universes and Springboard of Evolution

The birth of the first cell in primordial oceans, full of salty and boiling water some thirty-eight hundred million years ago was a big rise against the universal flow of entropy, which dictates an ever-increasing degree of disorderliness.

The powerhouse of the first prokaryotes, however primitive, took advantage of glucose as the main fuel to increase orderliness ingeniously.

Through anaerobic glycolysis by prokaryotes, two molecules of ATP were generated to maintain the integrity of the cell membrane and their primitive genetic material.

Plasticity of the genome, constant threat of the injurious environment, and acquired capability to mutate, built the foundation on which evolution continued its forward move.

The next chapter in evolution of living universes was encapsulation of the genetic material by another membrane and the birth of eukaryotes.

Thus, the living cell continued to build more barriers to protect its precious genetic material from the noxious agents of the surrounding harsh nonliving universe.

Meanwhile, a new powerhouse, namely aerobic glycolysis of nonoxidative phosphorylation subtype, was employed by both prokaryotes and eukaryotes.

As such, the living universes while developing another barrier to protect their precious genome from the harsh nonliving universe had built an intimate relationship with the resources of the surrounding nonliving environment.

Consequently, another precious resource, namely oxygen that existed in that harsh environment was extracted and taken into the buildup of their evolutionary fuel engine.

This could be considered a genius path, in the sense that precious elements, namely glucose and oxygen were extracted from the injurious environment, while more barriers were built for protection of the living cell from its harm.

The next step in evolution of living universes is one of the biggest puzzles in evolutionary biology.

The puzzle of acquiring mitochondrion by eukaryotes has continued to plague the field of evolutionary biology for several decades.

There are different hypotheses in this regard. One hypothesis is transfection of eukaryote with a bacterium called mitochondrion.

Another hypothesis is merging of two eukaryotes, one acting as the donor of mitochondrion and the other as recipient.

Both hypotheses admit to preexistence of mitochondrion as an independently evolved entity.

The possibility that mitochondrion could have come into existence through an independent path in evolution of living universes, is quite startling.

This implies that the outer membrane of mitochondrion came into existence with a dual function.

At one end, it acted as a barrier against the noxious non-living universe and protecting the inner membrane.

At the other end, it also acted as a participant and partner of the power-generating machinery of mitochondrion, namely the inner membrane.

This also strengthens the probability of coordinated evolution of outer and inner membrane of mitochondrion.

In addition, the mitochondrial circular DNA of around five thousand nucleotides does not have a definite membrane other than the inner membrane and is not protected by a separate barrier.

This is one of the most unusual paths in evolution of living universes.

Implications of this divergent path of evolution include a dedicated and separate path for the birth of a sophisticated energy generating machinery, namely mitochondrion.

As such, the outer and inner membranes of mitochondrion have adopted dual and divergent functions.

Accordingly, inner membrane is also acting both as a barrier protecting the mitochondrial genetic material as well as the powerhouse of mitochondrion.

This could be considered unusual in evolution of living universes, which has taken place by numerous dedicated steps, with each step not having more than one well defined task or category of tasks.

One example is the cell membrane, the main task of which is to act as a barrier for protection of the cell.

The pumps and receptors in the cell membrane act to secure the membrane integrity and serve the task of communication with the outside environment.

These subcompartment tasks are convergent in nature as far as the evolutionary role of that compartment is concerned.

In contrast, there is no such convergence of the dual tasks of the outer and inner mitochondrial membranes.

Thus, there seems to be a breakdown or dichotomy in the common theme of evolutionary steps as far as design and orchestration of function is concerned.

This is further complicated by the import of this structure into another living cell.

This would imply that the conventional design of evolution as well as its pace were dually ignored or insulted.

One would have expected that the eukaryotes without mitochondrion, would have acquired such a structure down the line through their natural history and conventional path of evolution of living universes.

However as mentioned in previous chapters, in quantum universe, there is nothing like a single path, and indeed we are dealing with sum over histories of innumerable paths from a broader perspective.

Thus, evolution of the quantum living universes is not bound by one carved in the stone path either.

In addition, because the whole process of evolution is happening in the infinity loop, such convergence of divergent paths could happen spontaneously to serve and promote the common purpose of evolution.

What makes the convergence of divergent paths in evolution of living universes more perplexing and exciting is further convergence of function of other separately evolved cellular machineries such as G-protein coupled receptors on the other divergently evolved and converged paths.

As such, one might consider convergence of function of sophisticated powerhouse of the cell, namely mitochondrion with G-protein coupled receptors as the main energy generator and sensors with their distal elements such as C-AMP as well as PI3 kinase and their downstream pathways, the springboard of evolution of living universes.

Through such collaboration and convergence of function of the members of this network, namely the sophisticated energy producer as well as energy sensor loops, balanced production, distribution, and storage of energy ensues.

Thus, the springboard for a forward move of evolution into and through multicellular era is born.

Mitochondrion also connects itself to apoptotic machinery and caspases through release of cytochrome C from its outer membrane during energy crisis.

This way, apoptosis would prevent catastrophic events such as malignant transformation, which could follow persistent decrease in cellular network free energy.

Breakdown of this mechanism is commonly seen in neoplastic disorders.

On a different note, the G-protein coupled receptors, meticulously and uninterruptedly communicate with the microenvironment at one end and intracellular compartments at the other end to maintain the energetics homeostasis of the living universes.

The landscape network energetics of G-PCRs acts as a bilateral relay station between microenvironment and cell.

The quantum features of G-PCR energetics landscape, which could act in a ligand dependent or independent fashion, would equilibrate energetics on both sides.

Accordingly, a low free energy microenvironment would lead to conformational change of G-PCR in such a way that GDP bound state would become dominant.

This would ultimately decrease intracellular C-AMP and activation of downstream pathways and oxidative phosphorylation induced excess ATP production.

Excess intracellular ATP, in turn would lead to generation and diffusion of humeral factors spilling into the adjacent microenvironment and increasing its free energy back to normal level.

This way, any deviation in network landscape energetics in any compartment would lead to appropriate corrective measures and restoration of normal state.

As such, the prerequisite for the birth and evolution of multicellular era, namely the development of an orchestrated and more sophisticated intra and intercellular, as well as microenvironment-cell communication network, which would also include the powerhouse of the cell is satisfied.

During this era, fine and well-balanced distribution of available free energy to members of the society of cells is of essence.

In addition, the need for controlled expansion of cell population becomes an evolutionary necessity.

In this sense, overpopulation is as much of a threat to the forward move of evolution, as is the inappropriate loss of crucial members of the population.

Hence, one could envision the birth and evolution of microenvironment as a fundamental step to secure this need through uninterrupted communication with the members of the society of cell population.

Microenvironment also acts as a medium, which secures balanced communication and inter-connectivity of the comprising cells of multicellular era with one another and outside world.

Thus, for the first time in the history of evolution of life, the concept of society gets carved in the matrix of living universes.

Physical cues are constantly recognized and relayed to the society of cells by the microenvironment.

This is reciprocated by the uninterrupted reflection of the energetics status of the cells onto a new microenvironment.

The connection of mitochondrion to Caspases through cytochrome C released from its outer membrane, which could activate programmed cell death machinery plays another brilliant role in this regard.

Thus, during energetics crisis, the powerhouse of the cell would sense the crisis and if needed could lead to demise of the cell in the interest of maintenance of homeostasis and survival of the cell population.

The evolution of multiple levels of inter and intracellular communication networks through microenvironment and major relay stations such as G-PCRs at one end and connection of mitochondrion to critical intracellular compartments, such as programmed cell death machinery at the other end, secures homeostasis of multicellular era.

Deep understanding of the multicellular era in evolution of living universes would enable us to

unravel the plaguing puzzles which are acting as
barriers for understanding and meaningful treatment of disorders such as cancer.

A significant number of catastrophic disorders of living universes, including malignancy have
their roots deep in the breakdown of foundation of
homeostasis of cellular energetics and interconnectivity of microenvironment with the society of cells
built during multicellular era.

Thus, the first and most important step in our
attempt at understanding of disorders, such as cancer, should be deep understanding of multicellular era and examination of the aberrancies that are
affecting these critical compartments.

Because of the intimate relationship between
cellular energetics and microenvironment, it is

plausible that dysfunction of one could spread to the other compartment instantaneously.

Therefore, it should not be surprising that corrective measures aiming at one could also lead to a shift toward normalcy of the other one.

Imids, such as Revlimid (Lenalidomide), commonly used in the treatment of malignancies such as multiple myeloma, interfere with growth promoting signals of microenvironment and augment the growth suppressing ones.

As of today, we are missing correlative studies examining the way mitochondrial biology and function is affected by these agents in patients who respond favorably to such treatment modalities.

In addition, there is urgent need for development of a new class of cancer therapeutics that would directly reverse the specific mitochondrial pathology in different malignancies.

Finally, analysis of G-PCR pathologies in different malignancies and targeted corrective treatment measures in this regard is another unmet and urgent need.

In closing the disruptions in the essential evolutionary triangle of G-PCRs, mitochondrion and microenvironment, which are the foundations of multicellular era, should be examined critically and urgently.

This should be done not only in malignant disorders, but also in a vast array of other disorders such as autoimmune and collagen vascular disorders.

Deep understanding of such aberrancies could open the window on a new era of game changing therapeutics.

CHAPTER 10 ADDITIONAL READING

Krishnan, A. 2015. Evolution of the G protein-coupled receptor signaling system. Genomic and phylogenetic analyses. Digital Comprehensive Summaries of Uppsala Dissertations from the Faculty of Medicine, 1116. 56.

Pfanner, N., Rassow, J., I. J. van der Klei, and W. Neupert. 1992. A dynamic model of the mitochondrial protein import machinery. Cell 68(2), 999–1,002.

Van Gestel, J., M. A. Nowak, and C. E. Tarnita. 2012. The Evolution of Cell-to-Cell Communication in a Sporulating Bacterium. PLoS Comput Biol8(12): e1002818

CHAPTER 11

Tumor Suppressor Genes, Proto-Oncogenes, and Telomeres, Guardians of Security Loop

The birth and evolution of living universes has happened in the salty boiling water of primordial ocean of the nonliving universe.

There are fundamental differences between these two universes as far as alignment with the most fundamental law prevailing the whole universe, namely the second law of thermodynamics is concerned.

The evolution of the known nonliving universe following its birth has happened spontaneously because of such alignment.

This is the case because the entropy or disorderliness of the expanding nonliving known universe has been increasing spontaneously following its birth.

On the contrary, the birth and evolution of living universes are based on constant struggle for generation and consumption of energy as well as surveillance and programming to minimize disorderliness.

During the unicellular era, the motive force for cell division was incessant penetration of proentropy forces through their cell membrane.

Generation of progeny with significantly less cellular network entropy, would thus secure the survival of unicellular organisms and their propagation.

As such, there is constant tug of war between the disorderly environment, which is empowered by the proentropy nature of the nonliving universe and the living universes, which are constantly trying to minimize the entropy that is spontaneously penetrating through their cell membranes and other built in barriers.

Thus, life could be considered a riot, in the sense that it is constantly fighting for generation and maintenance of orderliness, in a nonliving universe privileged with effortless and consistent increase in disorderliness or entropy.

Such riot demands constant generation and calculated consumption of energy to maintain the integrity of the living universes.

Heretofore, the limiting factor for the pace of cell division and survival in unicellular era has been the availability of free energy.

Cells that could not achieve energetics sophistication succumbed to proentropy environment and became extinct.

Consequently, capability of generation of free energy and its well-balanced consumption has been the prerequisite of survival of unicellular organisms.

As evolution of living universes moved from prokaryotes and eukaryotes to multicellular era, the need for more sophisticated genetic machineries to generate, store, and distribute energy became an evolutionary necessity.

During this era, it was not the survival of one single cell that mattered the most, rather the survival of the complex multicellular society became the main priority.

For this reason, genetic machineries for programmed cell death of single cells for the sake of survival and homeostasis of the society of multicellular organisms came into existence.

The need for generation, storage, and distribution of energy to the critical intracellular compartments of the cell at one end and the need for well-balanced proliferation and homeostasis of the cell population at the other end led to the evolutionary selection pressure for development of even more sophisticated genetic machineries.

In addition, overpopulation of cells became as detrimental to the survival of the society of cells as their underpopulation through inappropriate loss of cells.

Meanwhile, the incessant damage incurred by the surrounding environment, through numerous mechanisms and pathways such as ionizing radiation and oxidative agents, demanded evolution of a multitude of genes leading to formation of double and single strand DNA damage repair proteins and glutathione system among many others.

Thus, the cell had to safeguard, recover from, and survive the constant environmental damage and adjust its pace of mitosis not only for maintenance of its own integrity but also for maintenance of the integrity of the society of cells.

In a nutshell, in contradistinction to unicellular era in which the best interest of cell was in its own best interest, the best interest of a cell in multicellular era was reflected in the best interest of the society of cell population, as much as its own.

This demanded an ever-increasing pace of development of more complex genetic machineries, which would secure more sophisticated interconnectivity between and among cells at one end and among cells and microenvironment at the other end.

The development of intracellular communication network guided by cell surface receptors such as transmembrane tyrosine kinases, which could lead to activation of mitosis and their checkpoints as well as cyclin dependent kinase inhibitors, suppressing mitosis serve as some of the prominent examples in this regard.

The overexpression or overactivation of a gene whose product promote proliferation through a wide range of aberrancies such as point mutation is referred to as oncogene.

In contrast, loss of function of genes producing checkpoints and inhibitors of cyclin dependent kinases such as TP53 and RB referred to as tumor suppressor genes due to a wide range of aberrancies such as point mutation or deletion would lead to distorted and accelerated mitosis.

Thus, evolution of living universes has led to design of suppressors and activators of cell division to maintain homeostasis of the society of cell populations.

In addition, sharp radar-type sensors have come into existence to monitor and secure the safe number of cell division during the lifetime of a cell.

Tandem repeats of specific sequences of nucleotides at the end of each chromosome are referred to as telomeres.

Telomeres tend to shorten in length following numerous rounds of mitosis during the life span of each cell.

When the length of the shortened telomere reaches a critical level, the programmed cell death machinery of that cell gets activated and the cell undergoes apoptosis.

Under normal condition, the critical length of telomere at which apoptosis gets activated correlates with the degree of piled up DNA damage, which if unchecked and unabrogated, could significantly increase cellular network entropy and trigger malignant transformation.

Thus, telomeres and their control loops are acting as ingenious cellular machineries or radars, which secure protection of the cell through mechanisms such as apoptosis against emergence of catastrophic events such as malignant transformation.

In other words, the great design of telomeres of living universes leads to preemptive strikes, through mechanisms such as apoptosis to protect the integrity of the society of cell population.

As such, interconnectivity of the three major pillars of cell population control and homeostasis, namely tumor suppressor genes, proto-oncogenes, and telomeres is of essence.

Through widespread intracellular communication network, numerous mechanisms maintain homeostasis.

One example is freezing the cell cycle at G0 by TP53 following DNA damage, which allows time for repair of damage to get done.

Another example is PTEN mediated suppression of PI3 kinase, AKT and MAP kinase axis and the resultant cell proliferation signals.

Malignant transformation could happen, following a fundamental and irreversible dysfunction, or breakdown of communication between and among the guardians of the security loop, namely tumor suppressor genes, oncogenes, and telomeres.

Essentially, what guardians guard and secure is minimal allowable cellular network entropy, which correlates with maximum allowable free energy as per the limits set by the second law of thermodynamics.

Thus, however complex these interplays and cross talks seem to be, they are guided by this simple universal principle of living universes.

The reason for current failure of cancer therapeutics aiming at the members of this crucial triangle is lack of incorporation of this simple concept into the treatment strategy.

For example, telomere mediated cancer therapeutics is currently aiming at shortening of the long telomere of cancer cell while ignoring the underlying cause and effect principle.

In other words, ignoring the fact that the telomere of the cancer cell is inappropriately long because of the breakdown of the deeply seated mechanisms governing the homeostasis of the members of the triangle of guardians.

If the distortion irreversibly persists, it does not matter how capable we are at shortening the telomere length with our drugs.

This is akin to trimming the branches of a tree while leaving the root untouched.

As such, expectation of cure of cancer through current designs of telomere mediated therapeutics is more of an illusion.

Same is true with our current cancer therapeutics strategy dealing with deleted and mutated tumor suppressor genes and oncogenes.

There are three major possibilities that could potentially explain the dysfunction of each of the constituents of this crucial triangle are the following:

1. Inherent distortion in the structure or function of the constituent itself.

2. Distortion in communication network of constituents.

3. Combination of 1 and 2.

Consequently, the reason for long telomere in cancer cells is either inherent abnormality in telomere system, or distorted communication with tumor suppressor genes such as TP53 and RB, or oncogenes such as Ras, and finally a combination thereof.

The mechanism leading to dysfunction or mis-communication among the members of the triangle of guardians should be sought deep in the distortion of energetics governing the homeostasis of its members and their relationship.

This is simply because deep inside, the matrix of living universes is shaped and guided by homeostasis of their energetics.

As mentioned in previous chapters, Afrasiabi law of spontaneity dictates that at any given level of cellular network landscape energetics, the quaternary structure and orchestration of constituents of the cell from epigenome and DNA imprinting to microRNA network all the way to the protein-protein interactions change accordingly.

Double helix is not an exception. As such, different degrees of tension and relaxation in double helix could happen at different levels of cellular network free energy.

These physical changes would then lead to a change in their interactions. Such changes are diffuse and incorporate a broad spectrum.

The resultant changes in DNA imprinting, orchestration, and composition of microRNA network, RNA spliceosomes, RNA-binding proteins, and finally protein-protein interactions would seek restoration of minimal cellular network entropy.

As mentioned above, this represents the law of spontaneity. This law secures restoration of minimum allowable cellular network entropy, which correlates with maximum allowable free energy.

Accordingly, there is a constant tug of war proentropy and antientropy forces.

It is conceivable, that the system could fall prey to a breaking point which could affect one constituent or a combination of wide range of constituents.

Such could happen during extreme situations, which would not lend themselves to repair and reversibility.

Cancer and other catastrophic disorders are different manifestations of the breakdown of the law of spontaneity.

Consequently, any single pathology, such as long telomere in cancer cell, should get underpinned as to the root and source of the energetics aberrancies in that compartment or the closely related ones.

This kind of approach to complex biological systems would open a new window on understanding of disease and design of a new generation of cancer therapeutics.

The future of cancer therapeutics would rely on conversion of distorted energetics.

Such distortion is best defined as an irreversible drop in cellular network landscape free energy.

Such measurement is theoretically feasible through application of methodologies, such as liquid nuclear magnetic spectroscopy.

Conversion of cellular network entropy to its normal basal level could be designed by our future generation of quantum biologists.

Even today, we have many sophisticated tools in our tool box, such as nanodelivery and CRISPR technology as well as our knowledge of cellular sub-compartments such as microRNA and epigenome.

It is easy to imagine that the future generation of quantum biologists would identify the broken cellular node or nodes and would deliver nanomachines to those zones.

The nanomachines would restore the energetics homeostasis to the broken node or nodes through different biological and physical mechanisms.

Some theoretical examples, include modification of genetic or epigenetic code as well as

RNA spliceosomes, RNA-binding proteins, and microRNAs.

Modifications of distorted cellular energetics could theoretically also get done through regional delivery of appropriate and well calculated physical vibrations.

Delivery of nanovibrating machines to the diseased nodes in an attempt at increasing and thus restoring their free energy to baseline level could be one potential solution.

Such repair should preferentially be done in cells with inherent self-renewal capability or cells enabled accordingly.

Super quantum vectors as described in previous chapters could also get employed to modify the genetic programming of innumerable number of cells in tumor mass.

As such, we would start a new era and open a new window of opportunity on treating devastating disorders such as cancer.

CHAPTER 11 ADDITIONAL READING

Carracedo, A., and P. P. Pandolfi. 2008. The PTEN-PI3K pathway: of feedbacks and cross-talks. Oncogene, *27*, 5,527–5,541.

Lodish, H., A. Berk, S. L. Zipursky, et al. 2000. *Molecular Cell Biology.* 4th edition. New York: W. H. Freeman; Section 24.3, Oncogenic Mutations Affecting Cell Proliferation.

Rezaei-Tavirani, M., M. Rahmati-Roodsari, M. Mirzaie, P. A. Ggeram, and S. Sobhi. 2013. Cell survival entropy and cellular resistance activation dose: Effect of calprotectin on gastric adenocarcinoma cell line. Iran J Cancer Prev, Suppl, 12–16.

Martínez, P., and M. A. Blasco. 2017. Telomere-driven diseases and telomere-targeting therapies. The Journal of Cell Biology, 216(4), 875–887.

Mendoza, M. C., E. E. Er, and J. Blenis. 2011. The Ras-ERK and PI3K-mTOR Pathways: Cross-talk and Compensation. *Trends in Biochemical Sciences*, *36*(6), 320–328.

Michod, R. E. and D. Roze. 2001. Cooperation and conflict in the evolution of multicellularity. Heredity *86*, 1–7.

Pagliarini, R., W. Shao, and W. R. Sellers. 2015. Oncogene addiction: pathways of therapeutic response, resistance, and road maps toward a cure. EMBO Reports, 16(3), 280–296.

Sebé-Pedrós, A., B. M. Degnan, and I. Ruiz-Trillo. 2017. The origin of Metazoa: a unicellular perspective. Nature Reviews Genetics, *18*, 498–512.

Shcherbakov, V. 2005. Evolution as resistance to entropy. I. Mechanisms of species homeostasis. Zh Obshch biol. *66(3)*. 195-211.

Thompson, L. H. and C. L. Limoli. 2004. Origin, Recognition, Signaling and Repair of DNA Double-Strand Breaks in Mammalian Cells. In: Madame Curie Bioscience Database. Austin (TX): Landes Bioscience, 2000–2013.

Vicente-Dueñas, C., I. Romero-Camarero, C. Cobaleda, and I. Sánchez-García. 2013. Function of oncogenes in cancer development: a changing paradigm. *The EMBO Journal,* *32*(11), 1,502–1,513.

Webb, C. J., Y. Wu, and V. A. Zakian. 2013. DNA Repair at Telomeres: Keeping the Ends Intact. *Cold Spring Harbor Perspectives in Biology*, 5(6).

West, J., G. Bianconi, S. Severini, and A. E. Teschendorff. 2012. Differential network entropy reveals cancer system hallmarks. Scientific Reports *2(802)*, 1–8.

CHAPTER 12

The Past, Today, and the Future

Examination and analysis of the paths that we have gone through toward better understanding of living and nonliving universes and their implications in our everyday life is of essence.

A clear understanding of our position in the past and where we stand today would immensely help us in the design of our path toward future scientific goals in both realms.

LIVING UNIVERSES:

One way to look at the evolution of our understanding of living universes is in our current handling of the catastrophic diseases that affect our species, as compared with the past.

As has been rightfully said, cancer is the emperor of all maladies.

By looking back carefully at the last seventy-five years of modern history of cancer therapeutics, we would potentially develop the ability to envision and design the path to the future.

Treatment of the first case of lymphoma with nitrogen mustard in 1942, designed and performed by Karnofsky and Farber, gave birth to the new era of cancer (systematic cytotoxic) therapeutics.

In the last three quarters of a century, we have gone a long way. We have experienced both success and frustration in many ways.

Today, some seventy-five years later, we are probably at the beginning of a new path in the evolution of our thinking and design of future cancer therapeutics.

We started with single agent nitrogen mustard in 1940s. Fluorinated pyrimidines came on the scene in 1950s.

Multiagent chemotherapy, best exemplified by MOPP protocol designed by Devita for Hodgkin lymphoma came into existence in 1960s.

Serendipitous incorporation of cis-platinum into testicular cancer treatment protocols by

Einhorn revolutionized the treatment and outcome of germ cell tumors in 1970s.

Tamoxifen, as hormonal therapy and perhaps the first targeted therapy of cancer, became popular for treatment of hormone receptor positive breast cancer in 1980s.

The 1990s brought Taxol into treatment protocols of ovarian cancer, and later its extension into treatment protocols of lung and breast cancer.

At the same time introduction of monoclonal antibodies into the targeted treatment protocol of non-Hodgkin lymphoma, exemplifies another major advance of 1990s.

Specific targeted therapy against the driver translocation of chronic myelogenous leukemia, namely BCR-ABL, was the hallmark of advance in cancer therapeutics in 2000s.

The first seventeen years of the twenty-first century represents an explosion in targeted therapeutics of different kinds, including small molecules and monoclonal antibodies.

Tumor vaccines against papilloma virus, during the same time, is considered a major triumph against carcinoma of cervix.

The new generation of immunotherapy including checkpoint inhibitors, CART (chimeric antigen receptor T-cell), and BITE (bispecific T-cell engager) in the last few years have significantly improved the outcome of some deadly cancers.

Some examples include metastatic melanoma, refractory acute lymphocytic leukemia, and

non-Hodgkin lymphoma, non–small cell carcinoma of lung, and renal cell carcinoma.

We are now witnessing their extension into
other solid malignancies at a fast pace.

In summary, advances in our understanding of
cancer biology have paralleled advances and achieve-
ments in other fields of science, such as human
genome project and nanotechnology.

Most recently, CRISPR technology has empowered us with gene editing capability.

These discoveries and findings have led to the
design and implementation of more sophisticated
cancer therapeutics and significant prolongation of
cancer patients' lives.

We have continued to become more sophisticated cancer cell killers.

Most importantly, we have come to appreciate the limitations that we face as far as our expectations down this path of thinking and philosophy is concerned.

Conversion of cancer cell into normal cell by available treatment modalities is an exception rather than the rule.

This exception is best exemplified by treatment of acute promyelocytic leukemia with ATRA and arsenic trioxide.

In almost all the other cases, we have continued to remain obsessed with the theory and concept of killing the cancer cell as the golden gate to success in this field.

Lack of clear understanding of the mechanism of simple, daily events that we have taken for granted such as initiation of normal mitosis would keep us in dark regarding the deeply seated mechanisms involved in initiation and promotion of abnormal cell division, best exemplified by cancer.

As such, deep understanding of normal state is a prerequisite for basic understanding of cancer cell and ultimately cure of cancer.

In previous chapters, I have proposed the mechanism of initiation of normal mitosis, through interplay of the second law of thermodynamics with living universes.

The future of cancer therapeutics demands a radical shift in our thinking from killing the cancer cell to conversion of cancer cell to normal cell.

Furthermore, lack of deep understanding of chromosomal instability and the basic principles of intratumor heterogeneity and hyperplasticity of tumor cells comprising the tumor as well as the complex relationship of cancer cells with microenvironment would further complicate the task of cure of cancer.

Evolution of our thinking regarding cancer treatment design has brought us to a crossroad.

We either should shift toward a totally different strategy of conversion of cancer cell to normal cell or continue with more and more sophisticated design of cancer cell killing strategy.

History of cancer therapeutics evolution is the best attestation to the fact that expectation of cure is not realistic unless we change path.

There has also been a deeply seated element of ancestral instinct in our cancer treatment design, which dictates that killing the enemy is the best way to survive.

It is an evolutionary obligation to replace the concept of killing of cancer cell with the concept of conversion of cancer cell to normal cell.

As mentioned in earlier chapters, normal cell division is an attempt at decreasing the cellular network entropy to minimum allowable level.

In this regard, the limiting factor is dictated by the second law of thermodynamics.

Persistent cancer cell division represents a futile attempt at achieving that goal. This would paradoxically drive the system into a vicious cycle in neoplastic disorders.

As mentioned before, cancer cell network entropy has already escaped to infinity and is not subject to recovery.

To design conversion strategy in cancer therapeutics, we need to refer to available mathematical models and calculate master regulator complex entropy in cancer cell.

Combining the power of nanodelivery, CRISPR technology, and super quantum vectors could potentially enable us to achieve this goal. We could intervene at different levels, ranging from transcription machinery, epigenome, and microRNA network.

As such, we could redesign master regulator complexes with maximum allowable free energy, which correlates with minimum allowable network entropy.

This concept would act as the foundation and basic principle of conversion of neoplastic cell to its normal counterpart.

By employing the same fundamental principles, we could examine every single neoplastic disorder.

We could decipher their master regulator complex energetics barcode and design customized and tailored conversion measures.

Conversion could also target the driving factors behind chromosomal instability, which regulates mis-segregation and tumor heterogeneity.

Nanoconvertor machines should be designed in a way that they would be able to move one malignant cell to another.

This would be like the way viral infection spreads from one cell to another.

Consequently, at the end of the conversion cycle, all the malignant subpopulations would have got converted to their normal counterparts.

To achieve this goal, the quantum super vector should carry nanoenergetics sensors.

The exact level of cellular network entropy of the infected cell measured by quantum super vector would activate a conversion program specific to that energetic bar code.

As we travel to the future of living universes, not only we should master curing catastrophic disorders such as cancer, but also we should become able to design and clone new living universes.

Such living universes should live much longer and be capable of tolerating extremes, during migration to other planets and other corners of non-living universe.

THE PUZZLE OF MIND

Lastly, I would like to touch on mind and body interaction of living universes.

Mind continues to remain an ill-defined and not well understood entity with strong effect on living universes.

Among the most puzzling observations that I have had in the last many years with cancer patients is the miracle of hope and positive thinking.

For mysterious reasons, cancer patients of mine that have been able to maintain positive attitude and remain hopeful have continued to live much longer in comparison to records by literatures.

This holds true, stage by stage and disease by disease, while receiving the same treatment for their incurable malignancies.

What we refer to as mind might indeed be the conscious rays of cloud emitted from our neuronal network, surrounding, supervising, and controlling the function of the living universe from which it originates.

Indeed, cellular network energetics, might be under control of and subject to programming by the conscious cloud or the so-called mind.

We should look for a mechanism that connects positive thinking and hope to cellular network energetics.

One potential explanation is transmission of fine vibrations from the conscious cloud to the cells and modifying their frequency into one with lower network entropy.

As such, conversion of cellular network entropy could happen through the conscious cloud.

This could potentially affect all subcellular units and compartments, including genome, epigenome, microRNA network, and spliceosomes.

If we could identify the nature of conscious cloud or mind, we might become able to use it as an adjunct to cancer therapeutic modality.

NON-LIVING UNIVERSE

A similar critical examination and analysis of evolution of our understanding of the laws governing the nonliving universe is of essence.

Development of new concepts, which could lead to design of the future of space-time travel projects and migration out of this corner of universe is the next phase in our evolution in this regard.

We could consider our species, incarcerated in a tiny corner of nonliving universe now.

The walls of this prison are made of the invisible photons and the seemingly unsurpassable speed of light.

Without doubt, breakdown of this barrier would enable us to travel to undiscovered corners of the universe.

The discovery of gravity by Newton in seventeenth century could be considered the beginning of a new era in our understanding of the fundamental laws of non-living universe.

The next chapter was discovery of subatomic particles, such as electron, neutron, and proton, and later their constituents.

Discovery of photon as energy quantum by Max Planck was the next major event.

Deciphering the relation between mass and energy, and understanding the nature of space-time by Albert Einstein in early twentieth century, is another quantum leap in mankind's understanding of nonliving universe.

Heisenberg's uncertainty principle, Feynman's sum over histories, and Hawking's black hole theory as well as mathematical proof of big bang could be considered the most important additions to the discoveries of their predecessors.

In the last fifty or so years, a search for unification of quantum physics, which is based on uncertainty and relativity which is founded on certainty, has been going on.

Grand unified theory is supposed to reconcile and merge these two major concepts into one.

Our biggest challenge as living universes who are residents of this corner of nonliving universe is migration to other corners of nonliving universe.

This would allow the propagation of the most puzzling and miraculous gift, namely the gift of life and the gift of thinking to undiscovered corners of universe.

Lack of ability to achieve this goal would lead to annihilation of living universes off the face of nonliving universe.

Regular matter and regular energy are incarcerated in space-time, in a universe the fabric of which is made of dark matter and its quantum dark vibrations, namely dark energy.

For us as living universes to migrate to other corners of nonliving universe, we need to acquire the capability to merge into dark matter–dark energy compartment, which is perhaps what is known as wormhole.

Given the frail nature of living universes and their lack of ability to tolerate and survive such conditions, design of a new brand of living universes with such capabilities becomes an evolutionary necessity.

Alternatively, as we learn more about dark matter and dark energy, we might become able to design intelligent life made of those elements.

Such intelligent life does not have anything to do with the second law of thermodynamics at its core.

Its intelligence could conceptually be comprised of dark vibrating clouds, capable of perception, design, replication, and propagation.

We might down load intelligence and cognition into their dark vibrating clouds and let them act as our messengers to other undiscovered corners of nonliving universe.

As such, the future generations of living universes have massive amount of challenge threatening their survival heading their way.

Consequently, it would be prudent for living universes on this planet to set their differences aside and put their resources together to achieve the common goal of survival.

Living universes should be rest assured that by not doing so and not doing it urgently, they would follow the footprint of dinosaurs.

After all, as Stephen Hawking once said, it does not seem that the nonliving universe would care moving without us as the residents in its tiny, lost, dark, and cold corner.

CHAPTER 12 ADDITIONAL READING

Alpert, L. K., E. M. Greenspan, and S. S. Peterson. 1950. The treatment of the lymphomas and other neoplastic diseases with nitrogen mustard. Ann Intern Med 32, 393–432.

Burnett, A. K., N. H. Russell, R. K. Hills, D. Bowen, J. Kell, S. Knapper, Y. G. Morgan, J. Lok, A. Grech, G. Jones, A. Khwaja, L. Friis, M. F. McMullin, A. Hunter, R. E. Clark, and D. Grimwade. 2015. Arsenic trioxide and all-trans retinoic acid treatment for acute promyelocytic leukaemia in all risk groups (AML17): results of a randomised, controlled, phase 3 trial. Lancet Oncol. *16(13)*, 1,295–1,305.

Castle, P., and M. Maza. 2016. Prophylactic HPV vaccination: Past, present, and future. Epidemiology and Infection, 144(3), 449–468.

DeVita, V. T. Jr, A. A. Serpick, and P. P. Carbone. 1970. Combination chemotherapy in the treatment of advanced Hodgkin's disease. Ann Intern Med 73, 881–95.

Einhorn, L. H. 1990. Treatment of testicular cancer: a new and improved model. J Clin Oncol. 8(11), 1,777–1,781.

Gohil, S. H., S. R. Paredes-Moscosso, M. Harrasser, Vezzalini, M., A. Scarpa, E. Morris, A. M. Davidoff, C. Sorio, A. C. Nathwani, M. Della Peruta. 2017. An ROR1 bi-specific T-cell engager provides effective targeting and cytotoxicity against a range of solid tumors. Oncoimmunology, 6(7), e1326437.

Heidelberger, C., and F. J. Ansfield. 1963. Experimental and Clinical Use of Fluorinated Pyrimidines in Cancer Chemotherapy. Cancer Res 23, 1,226–1,243.

Johnson, L. A., and C. H. June. 2017. Driving gene-engineered T cell immunotherapy of cancer. Cell Research *27*, 38–58.

Jordan, V. C. 2014. Tamoxifen as the First Targeted Long Term Adjuvant Therapy for Breast Cancer. Endocrine-Related Cancer, 21(3), R235–R246.

Press, O. W., F. Appelbaum, J. A. Ledbetter, P. J. Martin, J. Zarling, P. Kidd, and E. D. Thomas. 1987. Monoclonal antibody 1F5 (anti-CD20) serotherapy of human B cell lymphomas. Blood, 69, 584–91.

Salasse, S., and C. M. Verfaillie. 2002. BCR/ABL: from molecular mechanisms of leukemia induction to treatment of chronic myelogenous leukemia. Oncogene *21*, 8,547–8,559.

Wood, A. J., E. K. Rowinsky, and R. C. Donehower. 1995. Paclitaxel (taxol). New England Journal of Medicine, 332, 1,004–1,014.

Kambiz Afrasiabi M.D. was born in Iran. He received an MD degree from Shiraz Medical School. Like many others, he left Iran and came to the USA to pursue his dreams. He completed his internship and residency in internal medicine at USC and his fellowship in cancer medicine from the National Cancer

Institute and UCLA. He has been involved in cancer research and practice since 1994. He has continued to do his independent studies about physical laws and their relationship with living cell. He has been curious about the origin of universe and life since early childhood. He believes that we need to change our path of thinking and approach to cancer in order to achieve cure. He holds an associate project scientist position at UCI. He lives with his family in Southern California.

www.ingramcontent.com/pod-product-compliance
Lightning Source LLC
Chambersburg PA
CBHW031124180526
45160CB00001B/17